The Solar System

Other Pergamon Titles of Interest

BARBATO
Atmospheres

GURZADYAN
Flare Stars

HEY
The Radio Universe, 3rd edition

KANDEL
Earth and Cosmos

KONDRATYEV & HUNT
Weather and Climate on Planets

SOLOMON & EDWARDS
Giant Molecular Clouds in the Galaxy

Pergamon Journals*

Chinese Astronomy and Astrophysics

Geochimica et Cosmochimica Acta

Journal of Atmospheric & Terrestrial Physics

Planetary and Space Sciences

Vistas in Astronomy

*Free specimen copies available on request

The Solar System

by

BARRIE WILLIAM JONES

The Open University, Milton Keynes, England

PERGAMON PRESS

OXFORD · NEW YORK · TORONTO · SYDNEY · PARIS · FRANKFURT

K.   Pergamon Press Ltd., Headington Hill Hall,
Oxford OX3 0BW, England

S.A.   Pergamon Press Inc., Maxwell House, Fairview
Park, Elmsford, New York 10523, U.S.A.

CANADA   Pergamon Press Canada Ltd., Suite 104,
150 Consumers Rd., Willowdale, Ontario
M2J 1P9, Canada

AUSTRALIA   Pergamon Press (Aust.) Pty. Ltd., P.O. Box 544,
Potts Point, N.S.W. 2011, Australia

FRANCE   Pergamon Press SARL, 24 rue des Ecoles,
75240 Paris, Cedex 05, France

FEDERAL REPUBLIC   Pergamon Press GmbH, Hammerweg 6,
OF GERMANY   D-6242 Kronberg-Taunus, Federal Republic of Germany

---

First edition 1984

Library of Congress Cataloging in Publication Data

Jones, Barrie William.
The Solar System.
Bibliography: p.
Includes index.
1. Solar system.   I. Title.
QB501.J66 1983   523.2   83–2165

British Library Cataloguing in Publication Data

Jones, Barrie William
The Solar System
1. Solar system
I. Title
523.2   QB501

ISBN 0–08–026496–4 (Hardcover)
ISBN 0–08–026495–6 (Flexicover)

*Printed and bound in Great Britain by
William Clowes Limited, Beccles and London*

To my family

PREFACE

The exploration of the solar system is an active and growing international enterprise. As our spacecraft range farther and farther from their home planet, and as the instruments that they carry become more elaborate and sophisticated, the number of fantastic discoveries has multiplied beyond many of the wildest fancies. More importantly, we have become much more aware of our planetary environment. Our Earth is just one of many bodies that circle the Sun. And like everything else in the universe, it has an intricate and fascinating history behind it, a history which it is important for us to decipher and understand. It has only been a century or so since geology began to talk seriously about the evolution of our Earth through long aeons of time. Now we are at the stage that planetary science can extend these discussions to other planets and to the solar system as a whole.

Four and a half billion years ago, our Earth began to accrete from planetesimals condensed out of the nebula of gas and dust that gave birth to our solar system. Much has happened since. For example, we now know that our atmosphere has changed slowly but significantly through time. One billion years ago, it had too little oxygen to support beings like ourselves. Has equilibrium been reached, or will our atmosphere change radically in the future? Fifty years ago, no one could have answered such a question. Today we have begun to try. Not only have we gathered the many facts that tell us what happened, but we are beginning to build adequate theories to tell us why and how. In part our increased understanding is coming from the study of our neighboring planets, Venus and Mars, both of which have also experienced profound changes in their atmospheric environments since their formation. From such comparative studies common patterns can be discerned, irrelevant idiosyncracies recognized, and general principles developed. Such knowledge is essential if we are to succeed in charting the future evolution of our atmosphere and of the changes that we are likely to induce by the ever-increasing stress that we impose on our environment.

Planetary exploration has documented many other examples of how things have changed and evolved in the solar system, and provided the many facts needed for understanding the physical, chemical, geological and atmospheric processes that govern such changes. We are in the process of learning in great detail how our solar system works today. We are beginning to decipher much about how it has worked in the past. The enterprise of applying this knowledge and understanding to projecting into the future must be considered one of the grandest intellectual adventures of all time. Every intelligent person has a right to participate in it, and a good introductory textbook is the first necessary step. Such a book will not answer all the questions, but it will provide the essential factual background and an introduction to the spirit of unrelentless reasoning needed to proceed further. The reader should not be satisfied with learning that our oceans may boil away five billion years from now, but should strive to understand why. It is one thing to observe that the tiny moon Phobos is spiralling into Mars, but quite another to figure out the process that makes this happen.

I hope that it is with this spirit that the reader will approach this introduction to the solar system. Facts are essential, but without understanding they are no more than a catalog. I also hope that this introduction to the subject will enable the reader to participate more fully in the continuing exploration of the sun, planets, and satellites, as well as in future investigations of comets and asteroids by spacecraft. The *Viking* landings on Mars and the *Voyager* flybys of Jupiter and Saturn are history. But right now as I write, two *Venera* spacecraft are on their way to Venus, probably to map the surface of the cloud-veiled planet by radar. *Voyager 2* is speeding to an encounter with Uranus in 1986. An international flotilla of spacecraft is being readied to intercept Halley's comet on its return to our neighborhood in 1986. What better time could there be to begin one's personal voyage of exploration of our planetary surroundings?

J. VEVERKA

Cornell University, Ithaca, NY
October 1983

ACKNOWLEDGEMENTS

I am very grateful to Geoff Brown for comments on drafts of Chapters 2 and 5, to an anonymous reviewer for comments on a draft of Chapter 3, and to Michael Woolfson for comments on a draft of Chapter 15. The manuscript was accurately and speedily typed by Bunty Beaugeard with considerable assistance from Sue Craig and Pam Taylor.

Laurie Melton helped me to track down several photographs and I am very grateful to Kris Acharia for preparing several glossy prints. Detailed acknowledgements for Figures are made near the end of this book.

CONTENTS

STUDY GUIDE

The main concern of this introductory textbook on the Solar System is with our present state of knowledge of the Solar System rather than with how that knowledge was acquired, though observational techniques and history are not entirely neglected. However, discussion of the Sun is limited to material relevant to the Solar System as a whole.

The book should be accessible to readers who come with very little background in science and mathematics. Such background as is necessary is included in the text, and whenever feasible it has been woven into the main story line. Otherwise, such background has been placed in sections marked ●. These sections interrupt the main story line, but you will need to read them as they occur unless the topics they introduce are already familiar to you.

An air of uncertainty pervades many parts of the story, but this is a true reflection of the present state of affairs.

When you first read this book you should do so from beginning to end in the order in which it is written omitting nothing and preferably attempting the questions at the end of each chapter. Thereafter it can be used as a reference text.

Cross-referencing

Chapters are divided into sections (e.g. 7.3) and subsections (e.g. 7.3.1), but for the purpose of cross-referencing within the text both are called *sections*. Any material preceding the first section of a chapter is cross-referenced as if it had been labelled with a zero (e.g. section 7.0), and any material preceding the first sub-section of a section is cross-referenced by the section number (e.g. section 6.3).

The Figures

These are an integral part of the text and therefore in many cases the captions are *not* intended to be complete.

Many of the pictures are taken from an oblique viewpoint and when this is not obvious then it is stated in the caption. In several pictures regular grids of black spots appear. These are grid marks *within* the cameras that took the pictures and are not a feature of the scene imaged.

The sources of the pictures and of certain drawings are given in the Figure acknowledgements.

The index

Entries in the index included *general terms* such as "ellipse", and areas of knowledge specific to particular astronomical objects, such as "Mars, present surface of". This second example shows that these kinds of entries are rather broad and that all the various multi-word terms are located in the index in accord with the initial letter of the *final* word. Multi-word general terms are treated in the same way. Definite and indefinite articles are of no significance, and *either* the singular *or* the plural is indexed, but not both.

The index includes people to whom reference is made in the text.

The *first* page entry in the index is where the *main* description occurs. Any further entries are where the description is significantly continued. The index does *not* include pages where the item is merely used or mentioned.

Sub-indexing occurs under most of the particular astronomical objects and under several of the general terms. In both cases the sub-indexing is a natural result of basing the index on the first letter of the last word in multi-word terms.

A developing subject

From time to time there are rapid developments in certain aspects of our knowledge of the Solar System, and a text which initially is up to date can be overtaken by events. This is happening already. Nevertheless, most of our basic ideas concerning the Solar System are not subject to rapid changes, and it is such ideas which form the greatest part of this text.

1

THE MAIN FEATURES OF THE SOLAR SYSTEM

Let's imagine that you and I have left this Earth and have travelled far into the depths of space. Let's also imagine that we look back to where we have come from and survey the Solar System from this distant vantage-point.

All around is the black brilliance of space, black except for the stars which shine brilliantly. But though our journey has shrunk the Sun to a point, it still outshines all the other stars. This is not because it is itself bright but because it is by far the closest star: though we are well beyond the outermost planet we are still only a small fraction of the distance from the Sun to its nearest stellar neighbour.

Out here we can see the Solar System as a whole. Yet at first sight this vantage-point doesn't seem to be all that good. The Sun is visible as a brilliant point of white light, but some of the planets are too dim to be seen and the rest are indistinguishable from faint stars.

What we need is a telescope. Fortunately we've brought one with us. So, let's turn in Sunwards and scan space for the planets. Now we can see them all, and all of them can be distinguished from the stars because each planet appears as a tiny disc of light. So too does the Sun. We can also see that the distances between the planets are vast compared with their sizes. In order from the Sun we see Mercury, Venus, the Earth, a ring of tiny planets called the Asteroids, then Jupiter, Saturn, Uranus, and, at almost the same distance, Neptune and Pluto. If you want to remember the order then here is a mnemonic to help you: "My Very Energetic Mother Always Jumps Superbly Under Normal Pressure." (I expect you can think of a better one, perhaps with astronomical connotations, ending with "Understanding Nine Planets" and somehow incorporating "Month" and "Mystery".)

Most of the planets have moons, that is, natural satellites going around them. The Earth has one, Mars two (both too tiny to be seen with our rather modest telescope), Jupiter sixteen, Saturn seventeen, Uranus five, Neptune three or four, and Pluto one.

Nearly all the mass in the Solar System resides in the Sun: that is, the Sun contains far more material than any other body. Then come the planets and the larger satellites, and then the Asteroids and the smaller satellites. Comparable in mass with the smaller Asteroids and satellites are the comets. Still less massive are the meteoroids which grade all the way through to tiny particles of interplanetary dust which thinly pervades interplanetary space. There is also a rarefied gas between the planets, most of which streams outward from the Sun, and is called the solar wind.

The Solar System does not end with the outer planets. Apart from the possibility of undiscovered planets beyond Pluto (which itself was only discovered in 1930) many of the comets have orbits which take them far beyond that planet, perhaps to a region where comets are comparatively common. And meteoroids, gas and dust are certainly present beyond Pluto too. The Solar System only ends where the gravity of nearby stars equals the gravity of the Sun, and that's something like half-way to those stars.

But now let's return to the planets and carry out a "thought experiment".

Imagine that there is a Cosmic Sheet of Paper, vast and flat, that we can manipulate in any way we please. I'm going to pass it through the centre of the Sun (it's non-flammable paper) and orientate it until all the planets are as close to it as possible. There. At once something rather striking is apparent: all the planets are very close to the paper. A sensible measure of closeness is the distance of a planet above (or below) the paper expressed as a fraction of the planet's distance from the Sun. At the moment, in 1982, Pluto is furthest from the paper: 30% of its distance from the Sun. Pluto, however, may have had a curious origin. For the other planets the distance is within 12% of their respective distances from the Sun.

Now it would be an extraordinary coincidence if we have carried this experiment out at just the one moment when all the planets and the Sun, are almost in the same plane subsequently to move to fill three-dimensional space more uniformly. Were we to stay here and observe the planets day by day, month by month, year by year, we would see that they *always* stay close to the plane of our cosmic paper. This is not a special time.

This remarkable coplanarity is a striking and important feature of the Solar System. And there are several others, as follows.

If we were to view our Cosmic Sheet of Paper from well above the North Pole of the Earth then we would see that all the planets move around the Sun in nearly circular paths in the *anti*clockwise direction. Were we to change our viewpoint to well below the South Pole the direction would be clockwise. Regardless of our viewpoint this direction is called the *prograde* direction. The opposite direction (clockwise from above the North Pole, anticlockwise from below the South Pole) is called the *retrograde* direction. Thus, the planets move in the prograde direction. Moreover, most of the satellites move around their parent planet in the prograde direction also. The path of one body around another is called its *orbit*.

As well as orbital motion the planets and satellites also *spin* around axes through their centres. The *North and South Poles* of the Earth are where the Earth's spin axis emerges through the Earth's surface. Most planets and satellites have spin axes that are roughly perpendicular to our Cosmic Sheet of Paper and in most cases the spin direction is prograde.

Thus, there exists in the Solar System a roughly circular swirl of motion, predominantly in the prograde direction, and approximately confined to one plane. Any theory of the origin and evolution of the Solar System must account for this striking feature.

Any such theory must also account for the "spread-outness" of the Solar System. The planetary orbits are *far* larger than the radius of the Sun. For example, the average distance of the Earth from the Sun is 215 times the solar radius, and in the case of Neptune this factor is 6463. Moreover, the Sun spins rather slowly, and therefore the rotational motion lies largely in the orbital motions of the planets.

There are also some curiosities which are striking, but the importance of which is not certain.

The best example is the *Titius-Bode Law*. This is a formula which generates the average distances of the planets from the Sun (more precisely, it generates the semi-major axes of the orbits — see section 1.1). There are several versions of the formula: I use the one illustrated in Figure 1.1. The

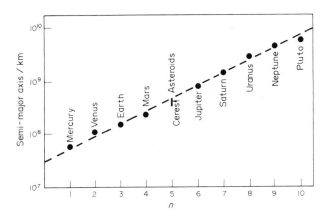

FIGURE 1.1
Semi-major axes of the planets' orbits and the Titius-Bode Law.

equi-spaced digits (n) on the horizontal (left to right) axis label planetary orbits. The distances on the "vertical" (bottom to top) axis are the semi-major axes of the corresponding orbits. The units are kilometres (km), written after the oblique on the axis label. The numbers are in *powers of 10 notation*, e.g. $10^7 = 10\ 000\ 000$. The *scale* on the vertical axis is *logarithmic*, that is, a given length along the axis, regardless of where it lies, equals a specific *multiple* of distance. For example, it is clear that a ratio of 10:1 always corresponds to the same length. The form of the Titius-Bode law that I am using here states that on such a graph the points representing planetary orbits will lie on a straight line. This means that the ratio of the semi-major axes of adjacent planetary orbits is the same for all pairs of adjacent orbits.

In Figure 1.1 the *actual* planetary orbits are shown as dots, and if they followed the Titius-Bode Law they would lie on the dashed line shown. Clearly the Law is not obeyed exactly. Moreover, if similar graphs are plotted for the orbits of large satellite families around their respective planets then a similarly modest degree of agreement is obtained.

The Titius-Bode Law was first proposed by the German scientist Johann Daniel Titius (1729–1796) in 1766, and, it seems, independently by the German astronomer Johann Elert Bode (1747–1826) in 1772. At that time the Asteroids were unknown, and therefore to the number 5 position in Figure 1.1 no known body was ascribed. It was left as a seemingly empty orbit in space because relocation of Jupiter and Saturn violated the Law. At that time Uranus, Neptune and Pluto were also unknown. Then, in 1781 Uranus was discovered, and it fitted the Law. In 1801 the first Asteroid was discovered, named Ceres, and it occupied an orbit close to that required at position 5. Neptune was discovered in 1846 and it too fitted the Law. However, by then several more Asteroids had been discovered, and the fit at position 5 looked less convincing. Today many hundreds of asteroids are known, and the line at position 5 in Figure 1.1 represents the range of semi-major axes of most of their orbits. In 1930 Pluto was discovered, and it does not fit the Law very well, though it may be that Pluto is an escaped satellite of Uranus or Neptune rather than a separately created planet.

It must be concluded that the status of the Titius-Bode Law is not clear.

Other important features of the Solar System include the sizes and compositions of the planets and the larger satellites. Figure 1.2 shows, to scale, the sizes of the larger bodies in the Solar System. This Figure tends to give an impression of the relative *areas* of the bodies, but the diversity of *volumes* is more marked. For example, the surface area of Jupiter is 120 times that of the Earth, but the volume of Jupiter is 1310 times that of the Earth. Figure 1.2 includes the Sun, the planets, the largest asteroid (Ceres), and the seven largest planetary satellites. On the basis of size alone it clearly makes sense to divide the bodies into three groups: first, the *giant planets* Jupiter and Saturn; second the *subgiant planets* Uranus and Neptune; third, all the rest.

In subsequent chapters you will learn that the composition of the bodies justifies the definition of giant and subgiant planets as distinct categories from each other and from the other bodies, but that the other bodies need to be divided into two groups, *rocky bodies* and *icy bodies*, terms which will become clear later. All of the numerous bodies too small to have been

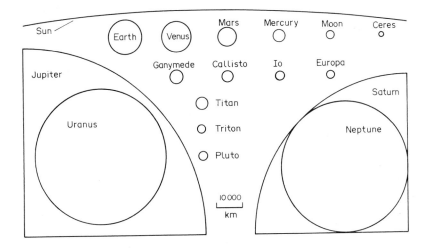

FIGURE 1.2
The major bodies in the Solar System (to scale).

included in Figure 1.2 fall into the rocky or icy categories. The larger of the rocky bodies, namely the Earth, Venus, Mars, Mercury and the Moon, are called the *terrestrial planets*.

The terrestrial planets occupy the inner zone of the Solar System, then come the giant planets and then the subgiant planets. This broad trend in composition and size must be accounted for by any theory of the origin and evolution of the Solar System.

The origin and the end of the Solar System and of any other planetary-systems is the subject of Chapter 15. In the intervening chapters the major bodies in the Solar System will be explored and their individual evolutions considered. I shall start with the Earth, a world that shares many attributes with the other terrestrial planets, but which in some ways is strikingly different. Most striking of all is that it is the only planet known for certain to bear life. From the Earth you shall journey to worlds that are increasingly alien. First to Mars, on which the surface conditions are more Earth-like than elsewhere in the Solar System, a surface of cool red dust under a thin clear atmosphere. Then to cloud-shrouded Venus, on which the surface conditions were once thought to be more Earth-like than on Mars, but which is now known to have a fiercely hot dusty surface beneath a massive corrosive atmosphere. Then to the Moon and Mercury, worlds devoid of atmosphere, with dusty cratered surfaces. And then to the giants and subgiants, extraordinarily alien worlds, though among their larger satellites we would feel, comparatively speaking, more at home. And then we shall visit Pluto, the comets, and the other small bodies that wander free of planets.

But before any of this it is necessary to examine orbits and axial spins in a bit more detail.

● 1.1 Orbits

The orbit of a body around the Sun is confined to a plane, and the Sun is in this plane, as illustrated in Figure 1.3 (a). The orbit is an *ellipse*, which can be

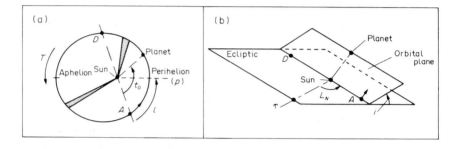

FIGURE 1.3
Orbital elements.

regarded as a circle viewed at an angle, the greater the angle the greater the departure from circular form. This departure is called the *eccentricity e* of the ellipse, and a circle can be regarded as an ellipse of zero eccentricity. The eccentricity of planetary orbits is very slight: the example in Figure 1.3 (a) is greatly exaggerated for clarity. The distance between the points marked "aphelion" and "perihelion" in Figure 1.3 (a) is the greatest distance between any two points on the ellipse. *Half* of this distance is called the *semi-major axis a* of the ellipse and I expect you can see that if an ellipse is regarded as a circle viewed at an angle then the semi-major axis equals the radius of this circle. The Sun is displaced from the centre of the ellipse along the semi-major axis to a particular point called the *focus* of the ellipse, the greater the eccentricity the greater the displacement. The body in orbit is closest to the Sun at *perihelion* (= "near to Sun") and is furthest from the Sun at *aphelion* (= "far from Sun"). The displacement of the focus is significant even when the eccentricity is small, and therefore it is possible to approximate an ellipse of low eccentricity by means of a circle centred on the focus. A second focus is equally displaced towards aphelion.

The plane of the Earth's orbit is called the *ecliptic*, and it acts as a reference plane for all other orbital planes in the Solar System. In this plane is chosen a reference direction γ, which is to be read "*the first point of Aries*" (a phrase I shall not explain). The direction γ is from the Sun to the Earth when the Earth is at that position in its orbit corresponding to the March equinox.

With respect to the distant stars the plane of the ecliptic and the direction of γ are (very nearly) fixed. The Earth's orbit also provides a convenient unit of distance in the Solar System, namely the *astronomical-unit*, equal to the semi-major axis of the orbit (1.496×10^8 km; km $\equiv$ kilometre $=$ 1000 metres (m)).

Figure 1.3 (b) shows the ecliptic and γ. It also shows the orbital plane of another body (in this case a planet). This plane intersects the ecliptic along the line *AD*. This line of intersection passes through the Sun, which lies in both planes. The angle L_N, measured *from* γ *to* the line of intersection in the prograde direction is called the *longitude of the ascending-node*. I don't want to pick through this term word by word, but the "ascending" means that the angle is taken to the line of intersection where the planet is moving *from* the South Pole side of the ecliptic *to* the North Pole side. This point on the planet's orbit is marked A in Figure 1.3 (b), the arrow showing the direction of planetary motion.

The angle *i* between the ecliptic plane and the other plane in Figure 1.3 (b) is called the *inclination* of the planet's orbit. Note that *i* is measured *from* the ecliptic *to* the other orbital plane in the space lying on the North Pole side of the ecliptic. Inclinations thus lie between 0° and 180° though all the planets have small inclinations the inclination in Figure 1.3 (b) being *a*typically large. Inclinations greater than 90° correspond to retrograde motion.

Now let's change our viewpoint and look face-on at this other planet's orbit. Let's also eliminate the reference plane, *except* for the line of intersection. The orbit in Figure 1.3 (a) will suffice, and the line of intersection crosses it at *A* and *D*. The angle *l from* the ascending node (*A*) *to* the perihelion in the direction of planetary motion is called the argument of the perihelion and is usually added to L_N to yield L_p the *longitude of perihelion*.

To specify where at any particular moment a body lies in its orbit it is necessary to know where it lay at some specified time in the past, such as at t_o in Figure 1.3 (a), and also the orbital period *T* which is the time taken by the body to move from any particular point on its orbit and next return there. The orbital period *T* is measured with respect to a vantage point fixed among the distant stars, and is called the *sidereal orbital period* ("sidereal" = "star related"). For the Earth this is *one year*.

Thus, the orbit of a body and its position in it are specified by the following *seven* items of information: the longitude of the ascending node; the orbital inclination; the longitude of perihelion (or the argument of perihelion); the semi-major axis; the eccentricity; the position at some specified time in the past; the sidereal orbital period. These seven items of information are called the *orbital elements*.

In an analogous manner the orbital elements of a satellite orbiting a planet can also be defined. Usually the ecliptic is replaced by the equatorial plane of the planet and γ by some other direction.

The manner of planetary motion was first established in the seventeenth century and is encapsulated in *Kepler's Laws of Planetary Motion*, named after the German astronomer Johannes Kepler (1571–1630). By the end of the century these laws had been explained with the aid of *Newton's Laws of Motion* and *Newton's Law of Gravity*, named after the British physicist Isaac Newton (1642–1727): these laws are very broad ranging and are not merely laws of planetary motion.

Kepler's first Law states that the orbit of each planet is an ellipse, with the Sun at one of the two focuses. The second Law states that equal areas of the ellipse are swept out by the planet in any fixed interval of time, as illustrated by the two equal shaded areas in Figure 1.3 (a). This means that the planet's speed around its orbit is greater the closer the planet is to the Sun. Kepler's third, and final, Law states that the *square* of sidereal orbital period divided by the *cube* of the semi-major axis has the same value for all the planets, and thus relates in a precise manner the semi-major axis to the sidereal orbital period.

Newton's Laws say that a planet moves because of the *gravitational force* by which any one mass will attract another, in this case the planet and the Sun. Imagine that the Sun and a planet are held fixed in space some distance apart by two Large Hands (the sort that write on Cosmic Sheets of Paper). The Hands release the Sun and the planet, and the gravitational force between them causes them to accelerate towards each other. This is called *free-fall*. The Sun is *far* more massive than the planet, and therefore the Sun hardly moves: I shall ignore its motion. Alas! the motion of the planet is short-lived, ending when the Sun and the planet collide.

Now imagine that before the Sun and the planet are released the planet is given some speed in a direction *other* than that pointing towards the Sun, as shown in Figure 1.4 (a). The planet accelerates towards the Sun but it no longer collides with the Sun but swings around it as shown in Figure 1.4 (b) deccelerates to aphelion and then accelerates towards its starting position as

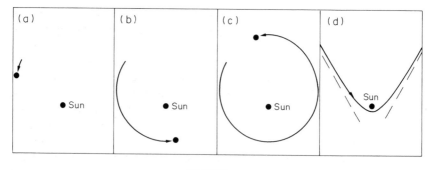

FIGURE 1.4
Orbital motion.

in Figure 1.4 (c). At its starting position it has its starting speed. This completes the first orbit of the planet around the Sun. It subsequently repeats its orbital motion. The orbit will be elliptical and its eccentricity will depend on the speed and direction of its initial motion.

Planets are not launched ready made in this manner, but the story illustrates that regardless of whether the Sun and planet collide an orbit is a state of free fall under the action of gravitational forces.

All of Kepler's Laws emerge from Newton's Laws, but Newton's Laws predict that another kind of orbit is possible, called a *hyperbolic orbit*. This is illustrated in Figure 1.4 (d). The body accelerates towards the Sun, misses, decelerates, but never stops moving outwards. It is lost to the Sun, and at a very large distance would move along the dashed line shown. If its motion were reversed it would go out along the other dashed line. The angle between these dashed lines measures the degree to which the orbit is hyperbolic. If the angle is small the orbit is called a *slightly hyperbolic orbit*.

The Large Hands could place a body in a hyperbolic orbit around the Sun by giving it sufficient speed when it is launched. In reality bodies which move in hyperbolic orbits through the Solar System have either come from interstellar space or they have been influenced by the gravitational force of a third body particularly Jupiter.

The gravitational influence on a body of bodies other than the Sun means that the orbital elements are not fixed. In the case of the planets they exhibit very slight, very slow changes. Among these changes can be discerned *periodic* changes, the elements slowly varying by small amounts around the average values. These average values are called the *mean orbital elements*.

The motion of a satellite or any other body around a planet is also explained by Newton's Laws in much the same way as the motion of the planet around the Sun is explained, the planet playing the role of the Sun. Though the Sun is far more massive than the planet the Sun is much further away and therefore the planet's gravitational field dominates the satellite motion, though the Sun can play the role of a third body. However, the mass of a satellite may not be negligible compared to the mass of the planet, and therefore the motion of the planet may not be negligible. Two bodies move in orbits around the *centre of mass* as illustrated in Figure 1.5 (a) and it is the centre of mass that lies at the common focus of the orbital ellipses. Figure 1.5 (b) illustrates the concept of centre of mass. Imagine that the planet and satellite in Figure 1.5 (a) are transported to an *enormous* planet on which a rod of negligible mass is used by Cosmic Hands to connect the two bodies and they are balanced on a knife-edge. The system will only balance if the knife-edge lies at the centre of mass. If the one body has negligible mass compared to the other then the centre of mass of the system lies at the centre of mass of the more massive body, and this is also where the focus lies. This is nearly the case for the Sun and planets.

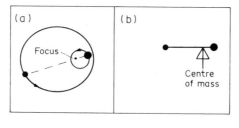

FIGURE 1.5
Centre of mass.

In addition to the sidereal orbital period there is also a more geocentric view of matters illustrated in Figure 1.6. The *synodic orbital period* of a planet is the time between certain alignments of the Earth, the Sun and the planet. For a planet in an orbit beyond the Earth's orbit, such as Mars in Figure 1.6, the alignment is with Mars at the point marked *opposition*. At opposition Mars, or any other planet beyond the Earth, is opposite in the sky to the Sun. The time between oppositions is the synodic orbital period of such a planet. For a planet such as Venus which lies within the Earth's orbit oppositions cannot occur and therefore the *inferior conjunction* (Figure 1.6) is used instead. Such planets also have *maximum elongations* east and west of the Sun (Figure 1.6) at which the angle between the planet and the Sun is greatest. All planets have *superior conjunctions* (Figure 1.6).

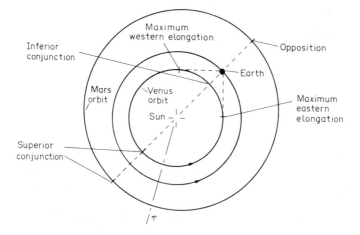

FIGURE 1.6
Various juxtapositions of planets. The viewpoint is perpendicular to the plane of the Earth's orbit, though the orbital inclinations of Venus and Mars are sufficiently small that their orbits would look little different were they viewed perpendicular to their own planes.

The *synodic* orbital period clearly differs from the *sidereal* orbital period. For example, consider a planet in an orbit well beyond the Earth's orbit. The sidereal orbital period will be many years, but the synodic orbital period will only be a little over the Earth's sidereal orbital period of one year, because in a year the distant planet does not move very far around its orbit and so it does not take long for the Earth to "catch it up".

Because planetary orbits are not perfectly circular, planets do not move at a constant speed around their orbits. This means that the time between oppositions, or between inferior conjunctions, is not quite the same from orbit to orbit, but undergoes small periodic variations around an average value called the *mean synodic orbital period*. The mean synodic orbital period is the same for all chosen alignments between the Earth, the Sun and a planet.

Table 1.1 lists, for the time of writing, two forthcoming oppositions for

TABLE 1.1
Oppositions, or maximum eastern elongations, of the major planets

Planet	Opposition, or maximum eastern elongation (ee)	Mean synodic orbital period/days
Mercury	21 April 1983 (ee)	115.9
	19 August 1983 (ee)	
Venus	16 June 1983 (ee)	583.9
	22 January 1985 (ee)	
Mars	11 May 1984	779.9
	10 July 1986	
Jupiter	27 May 1983	398.9
	29 June 1984	
Saturn	21 April 1983	378.1
	3 May 1984	
Uranus	29 May 1983	369.7
	1 June 1984	
Neptune	19 June 1983	367.5
	21 June 1984	

each of the planets outside the Earth's orbit. It also lists two forthcoming maximum eastern elongations for Venus and Mercury because at such elongations these two planets are most readily seen, in the Earth's western sky after sunset, though Mercury is elusive. Also listed are the mean synodic orbital periods of the planets, from which you can work out the approximate dates of subsequent oppositions or maximum eastern elongations.

At certain inferior conjunctions Venus and Mercury can be seen from the Earth to pass across the disc of the Sun. This is called a *solar transit*. These do not occur at every orbit because of the orbital inclinations. From Figure 1.3 (b) you can see that solar transits occur at those inferior conjunctions which occur when Venus or Mercury are near A or D. Table 1.2 lists some forthcoming solar transits and also some other important events that will occur over the next two decades or so.

TABLE 1.2
Some important forthcoming Solar System events

Event	Date	Event	Date
Solar transits of Mercury	13 November 1986 6 November 1993	Total solar eclipse	11 June 1983 22 November 1984 12 November 1985
Solar transits of Venus	7 June 2004 5 June 2012		3 October 1986 18 March 1988
Perihelion of Halley's comet	9 February 1986		

● 1.2 Axial spins

In addition to orbital motion each body in the Solar System spins around an axis that passes through its own centre of mass. The angle between the spin axis and the perpendicular to its orbital plane is called the *axial inclination* and for a planet orbiting the Sun is illustrated in Figure 1.7. (The

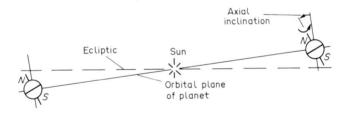

FIGURE 1.7
Axial inclination.

size of the planet is *greatly* exaggerated.) Note that the axial inclination is measured with respect to the orbital plane *of the body* concerned, and *not* with respect to the ecliptic except for the Sun the axial inclination of which is about 7° with respect to the ecliptic. Note also that, as the body moves around its orbit the direction of its spin axis does *not* change.

The axial inclinations of most of the planets and of most of the satellites are fairly small. Values less than 90° correspond to *prograde* axial rotation, whereas values between 90° and 180° correspond to retrograde rotation.

In just the same way that the sidereal orbital period is defined with respect to a vantage point among the distant stars the *sidereal axial period* T_a can be similarly defined.

The axial period of a planet can also be measured with respect to the Sun, as illustrated in Figure 1.8. This differs from the sidereal axial period because of the orbital motion of the planet around the Sun. The *solar axial period* is the time between successive passages of the Sun through some particular point in the sky, such as the noon position indicated by x in Figure 1.8. A solar axial period can also be defined in a similar way for a satellite.

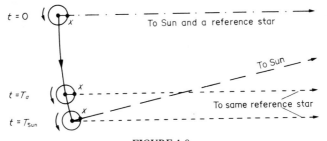

FIGURE 1.8
Sidereal and solar axial periods.

The solar axial period of a planet (or a satellite) varies with position in the orbit. The sidereal axial period is (very nearly) constant, but the ellipticity of a planet's orbit and the slight variation in speed as it moves around the orbit combine to cause periodic variations in the solar axial period around a mean value called the *mean solar axial period*.

In the case of the Earth the sidereal axial period (in 1982) is 23 hours (h), 56 minutes (m) and 4.0994 seconds (s). This is very nearly constant, currently increasing by about 0.0015 s per 100 years, a trend superimposed on shorter term but even smaller variations. The reason for the trend will be outlined in Chapter 5.

The mean solar axial period of the Earth is called the *mean solar day*, or the *day* for short. It consists of exactly 24 hours, and until recently this was how the *hour* was defined, as $1/24$ of the mean solar day. (It is now defined in terms of atomic systems that have nothing to do with planetary rotation.) The solar day can be up to about a quarter of an hour longer or shorter than the *mean solar day*. All our everyday clocks and watches keep time according to the mean solar day with the exception of sundials which record time according to the solar day.

The year consists of 365.2564 mean solar days, or 366.2564 sidereal axial periods. Note that there is exactly a difference of *one* between these numbers.

Table 1.3 summarizes the orbital elements and axial spins of the planets and also includes other information about the Solar System. Frequent

TABLE 1.3
Main properties of the planets

	terrestrial planets				
Property	Mercury	Venus	The Earth	The Moon[a]	Mars
Mean orbital elements					
Semi-major axis/AU[b]	0.387	0.723	1.000	—	1.524
Sidereal period/years	0.241	0.615	1.000	—	1.881
Eccentricity	0.2056	0.0068	0.0167	—	0.0934
Inclination/°	7.00	3.39	0.00	—	1.85
Long. of asc. node/°	47.1	75.8	—	—	48.8
Long. of perhelion/°	75.9	130.1	101.2	—	334.2
Properties of the planet					
Axial sidereal period/days	58.64	−243.01	0.9973	27.32	1.0260
Axial inclination/°	<3[h]	177.8°	23.45	6.68	23.98
Equatorial radius/km	2439	6052	6378	1738	3398
Mass/M_E[c]	0.0558	0.8150	1.0000	0.01230	0.1074
Mean density/(kg/m³)[j]	5430	5240	5520	3340	3940
Main constituents, volume fraction	rocky, 0.6 iron, 0.4	rocky, 0.85 iron, 0.15	rocky, 0.85 iron, 0.15	rocky, >0.99[i] iron, <0.01	rocky, >0.95 iron, <0.05
Atmospheric mass/(kg/m²)	negligible	1030 000	10 360	negligible	160
Mean surface temperature/K	440	730	288	250	218
Main atmospheric gases	—	CO_2	N_2,O_2	—	CO_2
Satellites and rings					
Number of major satellites[g]	0	0	1	—	0
Number of known minor satellites	0	0	0	—	2
Rings	none	none	none	none	none

(a) In this book the Moon is regarded as a planet. The Earth and the Moon share a common set of orbital elements, which is the orbit of the centre of mass of the Earth–Moon system. (b) 1 AU = 1.496×10^8 km. (c) $M_E = 5.98 \times 10^{24}$ kg. (d) Ceres is the largest asteroid. (e) At 1 bar level.

reference will be made to this Table.

In Table 1.3 no indication is explicitly given of the *precision* of the various numbers listed. For example, the mass of the Earth is given as 5.98×10^{24} kg (kg ≡ kilogramme). But the precision of the measurements which lead to this value is limited. A fuller statement is that the mass of the Earth *very* probably lies between 5.972×10^{24} kg and 5.980×10^{24} kg. Unless it is crucial to the argument I shall not normally refer to the precision of the values of quantities quoted in this book. A *rough* guide is that the last non-zero digit could change by 1 up or down. However, in some cases I have not quoted all the accurately known digits after a decimal point but rounded the value to the last digit given, and in some other cases trailing zeroes are *significant*, so that for example the equatorial radius of Saturn, given in Table 1.3 as 60 000 km, could *not* be as large as 70 000 km, nor as small as 50 000 km, but very probably lies between 59 900 km and 60 100 km.

TABLE 1.3
Main properties of the planets

Ceres[d]	giant planets		sub-giant planets		Pluto
	Jupiter	Saturn	Uranus	Neptune	
2.767	5.203	9.539	19.18	30.06	39.44
4.603	11.86	29.46	84.01	164.8	247.7
0.097	0.0485	0.0556	0.0472	0.0086	0.250
9.73	1.30	2.49	0.77	1.77	17.2
78.7	99.4	112.8	73.5	130.7	109.7
147.8	12.7	91.1	171.5	46.7	223
0.378	0.4135[f]	0.4440[f]	$-(0.5-1.0)$	$(0.4-0.8)$	6.387
<90	3.08	26.73	97.92	28.8	60(120)
410–570	71 400[e]	60 000[e]	25 900[e]	24 600[e]	1300–2000
$(1.87-2.07)\times10^{-4}$	317.9	95.15	14.54	17.23	0.0023
1200–3400	1340	705	1250	1640	410–1500
rocky, >0.99	H+He, >0.99	H+He, >0.98	H+He, 0.6 icy, 0.4	H+He, 0.5 icy, 0.5	icy, >0.5 rocky, <0.5
negligible	—	—	?	?	>20
170	—	—	?	?	50–60
—	—	—	H_2,He	H_2,He	CH_4,(Ar?)
—	5	8	5	2	1
—	11	9	0	1(2?)	0
—	small	very extensive	modest	?	?

(f) System III period. (g) mass greater than $M_E/10^6$. (h) "<" = less than. (i) ">" = greater than. (j) Below the solid surface for the terrestrial planets and Pluto, and below the 1 bar level for the others.

1.3 Questions

1. List at least six striking and important features of the Solar System.
2. According to the version of the Titius-Bode Law adopted here, where should a planet beyond Pluto lie? Give your answer in kilometres and in astronomical-units (AU).
3. Draw a sketch illustrating the seven orbital elements.
4. List Kepler's three Laws of Planetary Motion.
5. Two small bodies have equal mass and are in circular orbits around their centre of mass. Sketch these orbits, and discuss whether the bodies are both in free-fall.
6. What would be the sidereal orbital period of the new planet in question 2?

7. Describe, in a few words, the relationship between the sidereal and mean synodic orbital periods of a planet orbiting the Sun just beyond the Earth's orbit.

8. Using the data in this chapter calculate the approximate date of the opposition of Mars in 1988. Why is your answer not exact?

9. Discuss, in a few words, whether the mean synodic orbital period of a planet is the same regardless of whether opposition, superior conjunction or any other configuration is used to mark the beginning of such a period.

10. Why is there exactly a difference of *one* between the number of mean solar axial periods and the number of sidereal axial periods in one sidereal orbital period of the Earth? Would this difference of one also be the case for the other planets?

2
THE EARTH

To each of us the Earth seems vast. But in recent years the finiteness of our world has been made apparent by pictures taken from spacecraft, pictures like that in Figure 2.1. These have led to the term "Spaceship Earth" and to the belief that humanity, because it has become a global force, needs to write an operating manual for running the planet. Alas! the best operating manual that we could write at present would have some pages that would be very uncertain, others that would be incomplete, and some that would be wrong. Nevertheless, we are not entirely ignorant of our world nor of its workings.

FIGURE 2.1
The Earth imaged by Apollo 11 at a range of 180 000 km. North is near the top.

2.1 The Earth's interior

In attempting to establish the nature of the Earth's interior there are two difficulties to be faced.

First, the physical and chemical properties of candidate materials for Earth-building are not completely known, especially in the sort of conditions to be expected deep in the Earth, though there is some prospect that in the foreseeable future this difficulty can be largely overcome.

The second difficulty is more profound: our observations have to be made external to the Earth, or from relatively shallow depths in the Earth. Such observations do *not* provide sufficient information to enable the nature of the Earth's interior to be *deduced*. Instead, they can only *constrain* the range of plausible models of the interior as illustrated in Figure 2.2. Imagine that we have very few external observations. In this case the range of plausible models would be very large. As the number of external observations and their accuracy grows certain models would be ruled out whilst others would be considered very improbable (Figure 2.2). Ultimately the range of plausible models could be made very narrow though no unique model could emerge.

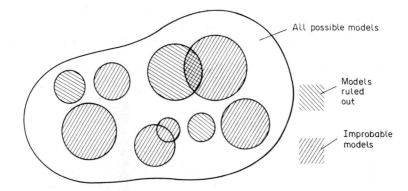

FIGURE 2.2
Observations and theories constrain the range of models.

I am going to present you with a model of the Earth, and then give you some idea of how well it stands up to being tested against various observational constraints.

But first, you need to know something about the constitution of matter.

● *2.1.1 The constitution of matter*

The basic "building block" of matter is the *atom*. A typical atom has a radius of about 10^{-10} m and consists of a cloud of moving *electrons* which account for most of the volume of the atom, and a tiny *nucleus* consisting of *protons* and *neutrons* and which lies near the centre of the electron cloud. But in spite of its small size the nucleus accounts for nearly all the mass of the atom. A neutron and a proton have about the same mass but differ in that the neutron is electrically neutral whereas the proton carries one unit of a type of electrical charge called *positive*. An electron, in spite of its far smaller mass, carries one unit of the other type of electrical charge, called *negative*. In an atom there is usually an equal number of electrons and protons and so overall the atom is electrically neutral. If this is not the case then the atom is called an *ion*. A neutral atom can become an ion through the gain or loss of electrons, a process called *ionization*. If an atom X has, for example, *lost* two electrons, then it would be denoted by X^{2+}. If it were to *gain* an electron then it would be denoted by X^-.

There are many different types of atom. The chemical properties are largely determined by the number of electrons, and you have seen that in the normal neutral state this equals the number of protons in the nucleus. The number of protons is called the *atomic-number Z*. A *chemical element* is an atom with a particular atomic-number. For example, oxygen has an atomic-number of 8. However, it is more usual to use a letter to denote an element rather than the atomic-number. Thus, the letter O denotes oxygen rather than 8. These letters are called *chemical symbols*. There are about 100 elements known, with Z ranging from 1 to about 100, with no gaps.

For any given element the number of neutrons in the nucleus, and therefore the mass of the atom, can vary. The number of protons plus the number of neutrons is called the *mass-number A*. An element with a particular mass number is called an *isotope*. Each element can exist as several different isotopes.

Table 2.1 lists the chemical elements of importance in this book. The mass numbers of the common isotopes are roughly double the atomic numbers shown, with the exception of hydrogen for which the common isotope has a

TABLE 2.1
Some chemical elements

Element	Z	Symbol	Element	Z	Symbol	Element	Z	Symbol
Hydrogen	1	H	Aluminium*	13	Al	Krypton	36	Kr
Helium	2	He	Silicon	14	Si	Rubidium*	37	Rb
Carbon	6	C	Sulphur	16	S	Strontium*	38	Sr
Nitrogen	7	N	Argon	18	Ar	Xenon	54	Xe
Oxygen	8	O	Potassium*	19	K	Lead*	82	Pb
Neon	10	Ne	Calcium*	20	Ca	Thorium*	90	Th
Sodium*	11	Na	Iron*	26	Fe	Uranium*	92	U
Magnesium*	12	Mg	Nickel*	28	Ni			

* Denotes a metallic element.

mass number of 1. The broadest division of the elements is into *metallic* and *non-metallic* elements. If a solid consists entirely or almost entirely of metallic elements then it will usually possess a particular set of properties among which will be those which correspond to the everyday characteristic properties of a *metal*, for example high lustre and malleability. All the characteristic properties of a metal arise from the release of electrons by the atoms of metallic elements when the atoms come close to each other, as in a solid. The metallic elements in Table 2.1 are asterisked.

A *molecule* consists of two or more atoms joined together. The *water molecule* is a familiar example and consists of two hydrogen atoms and one oxygen atom, represented as H_2O. An even simpler molecule is O_2. This is called *molecular oxygen*, and is the common form of oxygen in our atmosphere. The single atom form (O) is called *atomic oxygen*. A molecule is often called a *chemical compound* though this term also applies to a substance consisting of identical molecules. Such a substance is also called a *pure substance*. A substance in which not all molecules are identical is called a *mixture*. When a molecule exchanges one or more atoms with another molecule, or when a molecule gains or loses one or more atoms, then a *chemical reaction* is said to have occurred.

Table 2.2 lists the broad types of molecules that will be of importance

TABLE 2.2
Some types of molecule

Type	Essential element groups	Other elements
Oxides	O	almost any
Silicates	Si and O, e.g. SiO_4, SiO_3	usually metallic
Sulphides	S	usually metallic
Sulphates	SO_4	usually metallic
Carbonates	CO_3	usually metallic
Nitrates	NO_3	usually metallic
Nitrites	NO_2	usually metallic
Carbon compounds	C	almost any[+]
Hydrocarbons*	C and H	none
Hydrated compounds	H_2O	almost any
Hydroxides	OH	almost any

* This is a sub-type of carbon compounds.
[+] It is usual to exclude carbon oxides such as CO_2 and CO, and carbonates.

throughout this book. The first column is the name of the type, the second column gives the elements or groups of elements that characterize the type, and the third column gives the sort of elements that complete the molecule. Note that more than one of each essential element group can be present, for example, calcium hydroxide $Ca(OH)_2$ contains two *hydroxyl groups* (OH). A molecule can also contain essential element groups of various types, for example phlogophite, $KMg_3 (AlSi_3O_{10}) (OH)_2$, which contains hydroxyl groups and also silicate groups represented as Si_3O_{10}. It is an example of a *hydroxy silicate*. Phlogophite typifies the complexity of compounds that can occur in the Earth.

2.1.2 A model of the Earth's interior

Figure 2.3 shows the main features of the model. It consists of a specification versus depth of chemical composition, temperature and density, all of which I shall elaborate below. Note that there are variations *only* with depth and *not* with horizontal position along any surface at a fixed distance from the Earth's centre. This means that the model is *spherically symmetrical*.

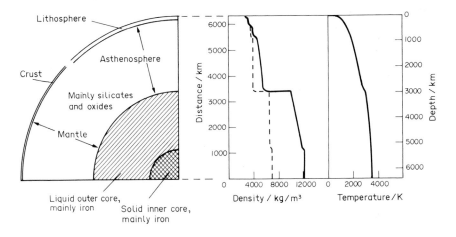

FIGURE 2.3
A model of the Earth.

The model divides the Earth into a *core* consisting largely of iron, an overlying *mantle* consisting largely of a mixture of various silicates and oxides, and an overlying thin *crust* which also consists largely of silicates though there are chemical differences from those which dominate the mantle. Magnesium is the most common metallic element in the silicates and oxides. There are also chemical variations within the mantle, but these are small and I shall not be concerned with them. More important chemical differences occur within the core and serve to divide it into an *inner* and *outer core*. These differences reside in the materials other than iron, and largely as a result of these chemical differences the inner core is solid whereas the outer core is liquid.

The mantle and crust are solid almost everywhere. But the "solidness" varies, and on this basis the Earth outside of the core can be divided as shown in Figure 2.3 into an asthenosphere and lithosphere as well as into a crust and mantle. The "solidness" of a solid can be measured by its *plasticity*. If you beat a metal with a hammer then permanent deformations can be caused. The greater the plasticity of the material the greater the ease with which this can be achieved. At comparable temperatures iron is less plastic than aluminium. But plasticity increases with temperature and near its melting point iron is fairly plastic as every blacksmith knows. Even concrete becomes moderately plastic at high temperatures. In the Earth, the crust and the uppermost part of the mantle have low plasticity and together form the *lithosphere*. Then there is a fairly sudden transition to much higher plasticity and this marks the top of the *asthenosphere*. The asthenosphere certainly extends to a considerable depth, perhaps nearly as far as the core.

Before any model is tested against external observations it has to be tested for *internal consistency*, that is, checks have to be made to see whether one part of the model contradicts another. In the case of the model in Figure 2.3 it is necessary to check whether the variations with depth of chemical composition, density, temperature, plasticity and the designation of a medium as solid or liquid, are consistent with each other. It is also necessary to check whether the different chemical compounds in contact in the interior can co-exist in the assumed proportions: chemical reactions can alter the assumed proportions.

These checks can only be properly made if we have laboratory data on the appropriate materials under appropriate conditions. Unfortunately, even at a depth of only 10% of the Earth's radius the pressures are already very high by laboratory standards and therefore there are many gaps in the laboratory data. Some of the gaps can be filled by theoretical models of how materials behave, but this still leaves areas of considerable uncertainty.

However, even though the model in Figure 2.3 cannot have its internal consistency thoroughly checked, it does stand up to such internal checks as can be made.

2.1.3 Global averages

The assumption of spherical symmetry may at first sight seem to be silly. Horizontal variations are all too evident to us, the ocean deeps and the Himalayas are two examples among many. However, the Earth's surface varies in height by only about 0.3% of the radius and therefore one must not be greatly impressed by the surface topography.

That the Earth as a whole is close to spherical symmetry is strongly indicated by many observations. For example, consider the variation of density inside the Earth.

The *density* at any point in a material is the corresponding mass of one unit of volume. Mass can be measured in a variety of units: I shall use the *kilogramme* which is abbreviated to *kg*. The unit of volume that I shall use is the *cubic metre*, which is abbreviated to m^3. Therefore, density will be measured in kilogrammes per cubic metre, written kg/m^3. For example, the density of the liquid water with which you are all familiar is about 1000 kg/m^3.

There are several external observations which indicate that the density in the Earth is close to spherical symmetry. For example, consider the *gravitational field* at the Earth's surface. The gravitational force between two bodies can be thought of as a *force field* surrounding the one body which acts on the other. If the Earth is regarded as the source of the field then the force on another body is proportional to the product of the strength of the field and the mass of the other body. Thus, measurement of this force yields the strength of the field. When such measurements are made at numerous points at the Earth's surface it is found that the field has very nearly the same strength at all of them. Moreover, the force everywhere points very nearly to the centre of mass of the Earth. It can be shown that this proves that the Earth's density is very close to spherical symmetry: in this particular case there are sufficient external data to make proof possible.

If the density is (close to) spherical symmetry then this strongly suggests that the chemical composition and temperature are also (close to) spherical symmetry, because composition, temperature and density at any depth in the Earth are closely related to each other.

Spherical symmetry is also expected on theoretical grounds, because the gravitational forces within a large mass tend to pull it into spherical form.

Thus, the model passes its first tests.

Note that surface gravity is not quite the same at all points on the Earth's surface. The broadest trend is a small but gradual *increase* in the force of gravity from equator to pole. This arises from the rotation of the Earth, which distorts it slightly from a spherical shape. The distortion is small, and has actually been included in Figure 2.3. It amounts to a mean polar radius 21.4 km *less* than the mean equatorial radius. The *polar flattening f* is the difference between these radii divided by the mean equatorial radius. In the case of the Earth the value of f is thus 3.355×10^{-3}.

The next test is chemical sampling of the Earth. Samples are acquired from mines and from bore holes drilled on land and in the ocean floor. Samples are also provided by natural processes such as volcanic eruptions though care must be taken to allow for chemical modifications that could have occurred between the source and the surface. Very little of the Earth can be directly sampled by these means: the deepest bore holes are about 15 km on dry land and about 2 km in the ocean floor, and natural processes bring up material from no more than about 100 km. These samples show that silicates account for nearly all the crust and upper mantle, and this feature has of course been incorporated in the model.

Further information about the composition of the Earth comes from its *mean density*. This is the *total* mass of the Earth divided by the *total* volume. The volume of the Earth can be obtained by means of various surveying methods, which I shall not describe. The mass of the Earth (M_E) can be obtained from the Earth gravitational field, which is proportional to its mass. The mass comes to 5.98×10^{24} kg. When divided by the volume this yields a mean density of 5520 kg/m^3, which is the same as the model. Let's see how the model achieves this agreement.

The greater the depth in the Earth the greater the pressure because of the growing depth of the overburden. When the pressure on a substance increases its atoms or molecules get pushed closer together and therefore its density increases. In Figure 2.3 the solid line on the density graph denotes the model densities in the Earth and the broken line denotes the lower uncompressed densities when the effect of the overburden pressure is removed (the corresponding change in volume is not shown). The mean density is then 4300 kg/m^3. Densities also depend on temperature, but this is a negligible effect compared to the effect of pressure in the Earth.

Table 2.3 lists the densities of some important materials. These values are at low pressure and can therefore be compared with the uncompressed densities in Figure 2.3. *Granite* and *basalt* in Table 2.3 typify the rocks that make up the crust. There is clearly a good agreement between the densities in the model crust and the densities of their typical crustal rocks. However, it is equally clear that these rocks cannot account for the whole Earth, because they are not dense enough. It is also clear that iron cannot account for the whole Earth beneath the crust, because iron is *too* dense. The model obtains

TABLE 2.3
The density of some important substances

Substance	density/(kg/m^3)[a]
Water	998
Granite[b]	2600
Basalt[b]	2900
Iron sulphide (FeS)	4840
Iron oxide (FeO)	5700
Iron	7900
Nickel	8900

Notes: (a) At $p = 1$ bar and $T = 273$ K.
      (b) These are complex substances of slightly variable composition with corresponding slight variations in density.

the correct mean density by adopting suitable proportions of the various materials that make up the core, the mantle and the crust.

But there is nothing unique about the recipe. The correct mean density could also be obtained with the same ingredients in different proportions and also with completely different ingredients except for the crust to which we have direct access.

Further constraints on the recipe emerge from a consideration of the materials that are likely to have been available to form the Earth. Crucial information is provided by the *solar relative abundances of the elements*. These are the observed relative numbers of atoms of each element in the Sun's atmosphere today. You shall see in section 15.2 that these relative numbers are highly likely to be very similar to those in the medium from which the whole Solar System formed, including the planets. I shall call this medium the *planetary formation medium (PFM)*, though it also includes the Sun.

These abundances help constrain models because it's no good trying to account for a large fraction of the mass of the Earth by means of materials entirely or largely made from elements of low abundance when materials made from more abundant elements will do just as well. For example, the elements iron and zinc have comparable densities and thus on the basis of density alone they are broadly interchangeable. But iron has a much higher abundance than zinc and therefore it is *far* more likely that the Earth contains large amounts of iron than of zinc.

Figure 2.4 presents the solar relative abundances of the fourteen most abundant elements, the volume of each cube being proportional to the number of atoms of the corresponding element. In the model of the Earth, oxygen, iron, silicon and magnesium are the four most abundant elements which between them account for over 80% of the Earth's mass and you can see from Figure 2.4 that all of these elements are in the top fourteen.

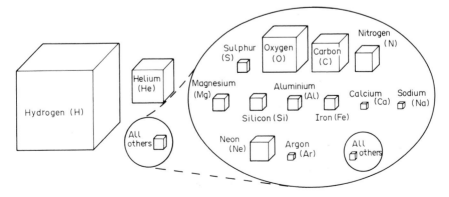

FIGURE 2.4
The solar relative abundances of the elements.

But this clearly can't be the whole story. For example, *hydrogen* is *far* more abundant than all the other elements *put together*, and yet in the model (not shown) it is largely confined to water (H_2O) in the oceans and in the atmosphere, and to hydroxides and hydrated molecules in the mantle and in the crust. Overall, hydrogen atoms probably account for no more than about 0.1% of the total number of atoms that make up the Earth. Moreover, the model cannot be wrong in this respect because there is plenty of evidence that hydrogen and its compounds cannot account for a large fraction of the Earth's mass: for example, water is not sufficiently dense, as you can see from Table 2.3.

The story lacks two crucial ingredients, and these are contained in the next two sections, on *phase-diagrams* and on the *accretion of planets*, topics which are of great importance throughout the book. In relation to the Earth they support the model feature that oxygen, iron, silicon and magnesium are the four most abundant elements.

Another indication of the composition of the Earth is provided by *meteorites*. These are relatively small lumps of rock collected by the Earth from interplanetary space, and which survive passage to the Earth's surface, where they are recovered. They are one of the main subjects in section 14.5 where you will meet a particular type of meteorite called the *carbonaceous chondrite*. In some respects these may represent the kind of material from which the Earth formed, and among the half dozen or so most abundant elements in such meteorites are oxygen, iron, silicon and magnesium.

● *2.1.4 Phase-diagrams*

Figure 2.5 shows the conditions under which a substance exists as a solid, liquid, or gas, that is, in which *phase* a substance exists. Figure 2.5 is an example of a *phase-diagram*.

You can see that the phase is determined by temperature and pressure.

The *temperature* of a substance is a measure of the random energy of motion of the molecules or atoms within it, the greater the energy of motion the higher the temperature. There are various scales for measuring temperature. I shall use the international scientific scale in which the temperature unit is the *Kelvin*, abbreviated K (not "degrees Kelvin", nor "°K"). The size of this unit is the same as that in the more familiar degrees centigrade scale (°C) but the Kelvin scale is offset by 273, so that, for example, 0°C becomes 273 K and 100°C becomes 373 K. The reason for this offset is that at 0 K the random energy of motion of the molecules or atoms in a substance is as small as it can possibly be. Therefore, temperatures below 0 K are not possible and 0 K is called *absolute-zero*. In degrees centigrade absolute zero is -273°C. The Kelvin scale is named after the British physicist Baron Kelvin of Largs (1824–1907).

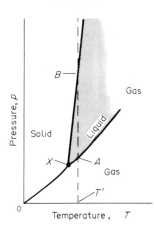

FIGURE 2.5
A typical phase-diagram of a stable pure substance.

It is important to distinguish between temperature and heat. *Heat* is energy which is transferred to or from a substance at a microscopic level and in a microscopically random manner. If heat is transferred to a substance then a common outcome is that the random energy of motion of the atoms or molecules in the substance is increased, and this is observed as a rise in temperature. If the substance *as a whole* is made to move, then though its overall energy of motion is increased the random motions inside it are not and there is no temperature rise.

The *pressure* of a substance is the force it exerts on its environment per unit area of its surface, and is the same in all directions. If the surface of the substance is neither expanding nor contracting then the environment must be exerting the same pressure on the substance. For example, the pressure of the Earth's atmosphere on your body equals the pressure exerted by your body on the atmosphere, otherwise the outcome would be messy. At sea-level the pressure of the atmosphere is very close to one *bar*, a unit of pressure which I shall adopt in this book. When you blow up a rubber balloon then on average, the pressure inside it will exceed the local atmospheric pressure by about 0.1 bar, the excess internal pressure being balanced by the stretch of the balloon fabric.

The phase-diagram in Figure 2.5 typifies the general features of the phase-diagram of most pure substances, that is, of substances which consist of just one chemical compound. It is also necessary that the compound is *chemically stable* over the pressure and temperature ranges shown, and that *thermal equilibrium* prevails, that is, the pressure and temperature of the substance are either constant or are changing only slowly. The phase-diagram of any particular substance will differ from that of another substance only in the detailed shape of the lines in Figure 2.5 and in the numerical values on the axes.

To get some feel for Figure 2.5 consider the temperature marked T'. At any given temperature the molecules in the substance have the same random energy of motion, regardless of the phase. At T' the substance is a gas at low pressures and has a comparatively low density. As the pressure increases, keeping the temperature fixed, the density increases. This is intuitively reasonable: an increase in pressure will tend to push the molecules closer together, and if the temperature doesn't change then there are no thermal expansion or contraction effects to complicate the picture. As the pressure crosses the *phase-boundary* between the gas and liquid phases at A in Figure 2.5 the substance turns from a gas into a liquid with a spectacular increase in density. The increase of density at B where the substance turns from a liquid into a solid is far less spectacular than at the gas–liquid phase-boundary. Within the solid region the density continues to increase with increasing pressure and there may be other phase-boundaries separating regions of the phase-diagram in which the molecules in the solid are arranged in different ways.

The relationship between density, pressure and temperature is called the *equation of state* of a substance. It provides one of the internal checks that have to be applied in Figure 2.3: the pressure at any depth can be calculated from the model, and thus the pressure, temperature and density of the substance at that depth are known and should agree with the equation of state obtained from laboratory measurements or from theory.

On the basis of its phase-diagram a substance can be placed on a scale of *volatility*, which is a measure of the range of pressures and temperatures over which a substance exists as a gas. The point marked X in Figure 2.5 is called the *triple-point*, because here all three phases meet. Broadly speaking the lower the temperature of the triple-point the more volatile the substance. Conversely, the higher the temperature of the triple-point the more *refractory* the substance. Substances with triple-point temperatures below about 500 K are called *volatiles*, whereas those with triple-point temperatures above about 1500 K are called *refractories*.

Water has a triple-point temperature of 273.16 K and is therefore a volatile, whereas iron, and many oxides and silicates, have triple-point temperatures in excess of 1500 K, and are therefore refractories.

The triple-point temperature of *molecular hydrogen* (H_2) is only 13.9 K, which makes it one of the most volatile substances known.

The more volatile a substance the more underabundant it is likely to be in a planet, as outlined in the next section.

When more than one type of molecule is present then the phase-diagram can be considerably more complicated than that shown in Figure 2.5. Furthermore, chemical reactions can occur between the different molecules.

Also, the molecules in a pure substance can become unstable over certain pressure and temperature ranges.

As well as determining the phase of a substance, pressure and temperature also determine the type of molecules yielded by chemical reactions. Broadly speaking, the higher the *temperature* the greater the extent to which a chemical reaction is favoured which increases the number of molecules present by generating a larger number of smaller molecules from a smaller number of larger molecules. This is because the random energies of motion increase with temperature and this tends to dissociate large molecules. *Pressure* has the opposite effect: in most cases the higher the pressure the smaller the number of molecules, in essence because as pressure increases the molecules are forced closer together.

● *2.1.5 Accretion of planets*

There is a broad consensus among most scientists that the planets were formed by a variety of processes the net effect of which was to bring together thinly dispersed materials, solid and gaseous, ultimately to build up the planet. Such a build up is called *accretion* (though some use this term in a more restricted sense).

In the PFM, depending mainly on local pressures and temperatures, some substances would have been present as solid particles but the more volatile substances would have been present as gases. In a gas the force of collision between one of its molecule and another particle of matter, such as another molecule or a solid particle, can be very large compared with the gravitational and chemical forces which would bind the molecule to these other particles and thus the molecule tends to rebound. By contrast solid (and liquid) particles in the same environment move much slower than individual molecules, and so the collisional forces are a lot less. This enables the gravitational and chemical forces to clump the solid particles together, and ultimately to form the planet, leaving the gases behind in what is becoming interplanetary space.

Gases are not entirely excluded. Small quantities will adhere to the *surfaces* of solids: this is called *adsorption*. Small quantities will also be trapped *inside* solids: this is called *occlusion*. Moreover, should a planet become sufficiently massive then it can gravitationally capture gases from interplanetary space, and in extreme cases this could eliminate any initial underabundance of volatiles.

In the case of the Earth, and of several of the other planets, substances which remained gaseous in the local PFM could not have been acquired in large quantities, and this readily accounts for the heavy depletion in such planets of hydrogen and of other elements associated with volatile substances. These planets thus become dominated by the non-volatile substances, such as iron and the various non-volatile oxides and silicates.

There are two extreme possibilities for the mode of accretion of a plant which can be outlined taking the Earth as an example. First, the Earth could have accreted its iron core first, and then added the mantle and crust. This is called *heterogeneous accretion*. Second, it could have formed as a homogeneous body with iron and mantle materials mixed throughout its volume. This is called *homogeneous accretion*. In this case the core must clearly have somehow separated later. This separation is called *differentiation*. A body which is heterogeneous, regardless of whether it underwent differentiation, is said to be *differentiated*.

2.1.6 Seismic evidence

A powerful test of the depth variations in the model is provided by waves which travel through the Earth, called *seismic waves*. These are examples of a broader class of waves called *elastic waves* in which the wave is a to-and-fro oscillation of matter. Seismic waves can be very destructive when they appear at the Earth's surface, where they constitute *Earthquakes*.

A major natural source of seismic waves is strains which build up in mantle and crustal rocks in response to various stresses. If the rock suddenly yields then seismic waves are generated. This is rather like plucking a string. Seismic waves can also be generated by natural impacts on the Earth's surface, and by natural and artificial surface explosions.

There are several different types of seismic wave, some of which can only exist at or near abrupt changes of density, such as the surface of a planet. But it is mainly the other types, which pass through the Earth, that have been of particular value in elucidating the structure of the Earth's interior: in particular, P-type waves and S-type waves. These types can be distinguished from other types and from each other by careful analysis of the seismic wave records obtained at the Earth's surface. Such records are obtained by instruments called *seismometers*.

In *P waves* the to-and-fro motion experienced by the material through which the wave is passing is along the same direction as that in which the P wave is travelling, as shown in Figure 2.6 (a). *In S waves* the to-and-fro motion is at *right angles* to the direction in which the S wave is travelling, as shown in Figure 2.6 (b). Whereas P waves can travel through material in any phase, S waves, for all practical purposes in the Earth, can only travel through solids. Liquids (and gases) are too readily sheared to sustain S waves to an appreciable extent.

A particular P wave or S wave does not carry separable information about the materials at each and every point on its path, but only the *combined* effects of the materials at all of these points. It was therefore no easy matter to build up a picture of the Earth's interior from the seismic records. But gradually,

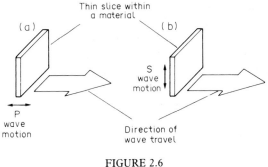

FIGURE 2.6
P waves and S waves.

from beginnings at about the turn of the century, and by means of arrays of seismometers all over the globe, a seismic picture of the Earth's interior has emerged.

This picture is shown in Figure 2.7, and is the variation with depth of the speeds at which the P and S waves travel. These speeds do not seem to vary much with horizontal position, except in the lithosphere, and this provides additional support for spherical symmetry.

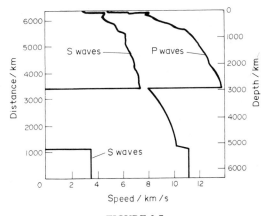

FIGURE 2.7
P-wave and S-wave speeds in the Earth.

The P and S wave speeds in a material depend on the density of the material and on two independent *elastic-constants*. These constants define the extent to which the material is temporarily deformed whilst it is subject to various stresses. (Do not confuse this with the permanent deformation associated with plasticity.) Thus, at each point in the Earth there are *three* unknown material properties but only *two* relevant observations, namely the P and S wave speeds. Therefore, it is not possible to deduce from these speeds the density and the two elastic-constants. Moreover, even if the density and the elastic-constants were known there would probably be many materials that would fit the values.

In recent years the way around this seeming impasse has been to bypass the data in Figure 2.7 as follows. First the densities and the P and S wave speeds at various depths are *guessed*. Then these data are used to calculate certain observed quantities, namely: the mass of the Earth and its polar moment of inertia (see section 3.2.1) both of which depend on the density versus depth; the travel times from source to seismometer of selected P and S waves, which depend on the density *and* on the speeds versus depth; and the oscillation periods of certain *free oscillations* of the Earth, which also depend on the density and the speed versus depth. These free oscillations are elastic waves on such a large scale that the whole Earth can be considered to be flexing in various ways. If the guessed depth variations of density and P and S wave speeds do not yield values of these other quantities sufficiently close to the observed values then the guesses are changed until they do. This type of approach is called the *Monte-Carlo Method*, though luck plays a smaller role than at the gaming table.

The outcome of this method is that the density versus depth in the Earth is strongly constrained. However, the elastic constants are less strongly constrained, which allows some freedom in modelling the composition at various depths. The seismic data also provide some constraints on temperatures because the elastic-constants vary a bit with temperature.

The model in Figure 2.3 meets the constraints imposed by this seismic (and related) data.

There are several features in Figure 2.7 to which I draw your attention. First, at roughly 50 km below the surface of the Earth the P and S wave speeds suddenly increase by a large amount. This points to a compositional change and corresponds to the crust–mantle interface in Figure 2.3.

Second, just below this interface, particularly beneath the oceans, the wave speeds *decrease* a little with increasing depth, yielding a narrow region of lowered speeds called the *low velocity zone*. It is widely believed that in this zone, which lies within the asthenosphere, a few percent of the material present is liquid. Such a mixture of solid and liquid phases can readily arise from a mixture of substances because at any given pressure a mixture melts over a *range* of temperatures and therefore at a particular temperature only a fraction of the material can be molten, more of some components than of others.

Third, any plasticity of the asthenosphere does not prevent S waves travelling through it. The S waves clearly show that the mantle is largely solid, but the asthenosphere could still be appreciably plastic as far as *sustained* stresses are concerned and still respond to the comparatively *rapid* to-and-fro motions of S waves.

Fourth, S waves do not traverse the depth range corresponding to the outer core. This *strongly* suggests that the outer core is liquid. The decrease in the P wave speed at a depth corresponding to the core–mantle boundary is too large to be explained by a change of phase alone, and therefore there is also a change in composition at this depth, which in the model is a change from mainly silicates and oxides to mainly iron. The slight increase in the P wave speed at a depth corresponding to the inner core–outer core interface is also too large to be explained by a change of phase alone. The seismic data are consistent with the predominance of iron in both parts of the core but suggest that the remaining components are different. For example, in the model in Figure 2.3 the mass fractions (not shown) in the inner core are 80% iron, 20% nickel, and in the outer core are 86% iron, 2% nickel, 12% sulphur. Other models incorporate silicon, oxygen, carbon, chromium and titanium, with various amounts in each part of the core.

The solidification of the inner core is *not* because of lower temperatures there: in the model in Figure 2.3, and in the actual Earth, it seems certain that there is no such decrease. It is largely because of the composition change, with a small measure of assistance from the increased pressure.

2.1.7 The magnetic field and the Earth's core

A *magnetic field* is another example of a force field, the gravitational field having been introduced in section 2.1.3. Figure 2.8 (a) illustrates the Earth's gravitational field, the arrows showing the direction of the field and the closeness of the lines indicating its strength. You can see that the gravitational field points towards the centre of mass of the Earth, this being

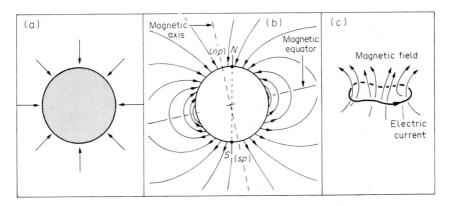

FIGURE 2.8
(a) The Earth's gravitational field. (b) The Earth's magnetic field. N and S denote the north and south poles of the spin axis and np and sp those of the magnetic axis. (c) A magnetic dipole source.

the direction of the force of the Earth on another body, and that the field strength increases as the distance above the Earth's surface decreases.

The magnetic field is a different type of field altogether.

Figure 2.8 (b) illustrates the *Earth's magnetic field*. You can see that it has a different spatial form from the Earth's gravitational field. This particular form of magnetic field is called a *magnetic dipole field*, and this is the most common form of magnetic field. It decreases with distance more rapidly than the Earth's gravitational field.

The *magnetic axis* in Figure 2.8 (b) is a line around which, to a high degree of accuracy, the field lines can be rotated to form shells which represent the field in three dimensions. The lines in Figure 2.8 (b) can be regarded as sections which divide these shells into equal halves. If the direction of the arrows on the lines is ignored then the shells are also divided into equal halves by the *magnetic equator*, seen from the edge in Figure 2.8 (b). The point at which the magnetic axis passes through the magnetic equator is the *centre* of the magnetic field. The points marked *np* and *sp* in Figure 2.8 (b) are, respectively, the *north magnetic pole* and the *south magnetic pole*, and are the locations on the Earth's surface where the magnetic axis passes through. If the magnetic field had the opposite direction then the arrows on the lines would point in the opposite direction and the labels *np* and *sp* would be exchanged. The direction of the field lines defines the *polarity* of the field.

The line *NS* in Figure 2.8 (b) is the spin axis of the Earth, and you can see that the magnetic axis is inclined several degrees from the spin axis. You can also see that the magnetic field centre does not lie at the centre of the Earth. Indeed, with the plane of this diagram defined to include the spin axis, the magnetic field centre, and also *sp*, lie a little way behind the plane, and therefore the magnetic equator is not quite viewed edge-on.

The strength of magnetic fields is measured in a variety of units. I shall use the *gamma*. The magnetic field at the Earth's surface is greatest at the poles, where it is about 60 000 gamma, and is least at the magnetic equator, where it is about 30 000 gamma.

The term "dipole field" arises from a now obsolete view of the source of such fields. The modern view is represented in Figure 2.8 (c). The arrow on the closed loop represents the direction of an electric current. An *electric current* is a flow of electric charges, such as electrons or ions. A material which readily allows the flow of currents is called an *electrical conductor*, and a measure of this readiness is its *electrical conductivity*. The corresponding quantity for heat transfer is the *thermal conductivity*. An electric current generates a magnetic field, and if the current forms a closed loop as in Figure 2.8 (c) then it is called a *magnetic dipole source*. At a distance from this source that is large compared with the source's size the field is a dipole field as in Figure 2.8 (b) and its spatial form is independent of the shape of the loop. Closer to the loop the spatial form changes, and becomes dependent on the loop's shape. The pattern of current flow can be very complicated, consisting of many intertwining loops, but as long as there is a net current in a closed path then the source constitutes a magnetic dipole source. The strength of a dipole source is measured by its *magnetic dipole moment*. The greater the current and the greater the area covered by the current loops the greater the magnetic dipole moment. At any given distance from a magnetic dipole source the magnetic field strength is proportional to the magnetic dipole moment.

You are probably familiar with iron magnets. Here too the field comes from currents in motion. There is a myriad of dipole sources of atomic dimensions which yield fields that add up to give a powerful external field.

The best theory of dipole sources inside planets, including the Earth, is the *dynamo theory*. A dynamo is a device which converts energy of motion into energy to drive electric currents. In the case of a planet an important energy of motion is its orbital and axial rotation. Another is *thermal convection*, which is the motion of liquids driven by heat and which will be described in section 2.1.12. If there is a *liquid* electrical conductor in the planet then rotational and convectional motion can give rise to electric currents in the liquid, and these currents can constitute a dipole source: a *solid* conductor will not do. In the beginning it is probably also necessary for there to be an external magnetic field which need not be strong but which enables the external source to become established. Even today there are sufficiently powerful fields external to the planets, particularly the field of the Sun. Once a planetary field is established these external fields are not necessary to sustain it. The planetary field is expected to have a small but non-zero inclination with respect to the spin axis of the planet, and to have a field centre slightly offset from the planet's centre of mass.

The liquid core in the Earth model in Figure 2.3 satisfies the constraints imposed by the observed magnetic field and by the dynamo theory. Moreover, the slow growth of the inner core by the segregation into it of the appropriate materials from the outer core could provide sufficient heat of differentiation (section 2.1.14) to drive thermal convection in the outer core and hence sustain the electrical currents. It is also possible that radioactive decay (section 2.1.14) is a major source of heat in the core.

A rather puzzling feature of the Earth's magnetic field is the abundant evidence which shows that the field has undergone many reversals of direction in the past. Less dramatically, the direction of the magnetic spin axis is continually changing by small amounts and so too is the field strength. The dynamo theory is not sufficiently well understood to account in detail for such changes, though it does seem able to account for the main features of the Earth's field.

2.1.8 The Earth's lithosphere

You will recall from Figure 2.3 that there are two boundaries in the outer regions of the Earth and that they are of two different kinds: the crust–mantle division marks a compositional change whereas the lithosphere–asthenosphere division marks a change in plasticity. The lithosphere includes all of the crust and the uppermost part of the mantle.

You will also recall that the crust–mantle division is apparent in seismic data (Figure 2.7) and indeed it was from seismic data that this division was discovered in 1909 by the Yugoslavian geophysicist Andrija Mohorovičić (1857–1936; ". . . čić" is pronounced "chick"). The interface between the crust and the mantle is appropriately named the *Moho*. It marks a transition from lower density silicates in the crust to higher density silicates in the upper mantle, the transition occupying a depth range of 0.1 to 0.5 km. The Moho lies between about 5 and 10 km beneath the ocean floors and between about 20 and 90 km beneath the surfaces of continents, the depth of the Moho varying from place to place.

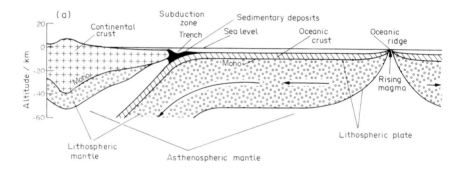

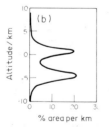

FIGURE 2.9

(a) A simplified cross-section of a typical portion of the Earth's lithosphere. The vertical scale is exaggerated. (b) The hypsometric distribution for the Earth. This shows the percentage of the surface that lies within each kilometre of altitude (0 km = sea-level).

The asthenosphere is more apparent in gravitational studies than in seismic studies as you shall see in section 2.1.10, and gravitational studies also reveal the crust–mantle division though I shall not go into this. Chemical analyses of the different types of rocks (section 2.1.9) and their distribution across the surface of the Earth also reveal the crust–mantle division, and moreover show that there are different types of crust, and this is also revealed by seismic and by other studies.

Broadly speaking, the Earth's crust can be divided into two types called *continental crust* and *oceanic crust*. These largely consist of slightly different types of silicates. The two types of crust are illustrated in Figure 2.9 (a). Note that the oceanic crust does not extend *beneath* the continental crust. Note also that though continental crust reaches higher altitudes than oceanic crust and thus accounts for most of the dry land, with oceanic crust lying above sea level in only a few places such as certain oceanic islands, the boundary between the two types of crust is *not* the sea shore: continental crust in most regions only gives way to oceanic crust at considerable depth as you can see in Figure 2.9 (a). Only if the volume of water in the oceans was considerably reduced would the sea shore roughly correspond to the transition between the two crustal types.

Continental crust not only stands higher on average than oceanic crust, but also lies at largely *separate* altitudes as illustrated in Figure 2.9 (b). This is called a *hypsometric distribution*, and you can see that the distribution by area of the various altitudes of the Earth's surface forms two peaks. The higher peak corresponds largely to continental crust and the lower peak largely to oceanic crust.

2.1.9 Rocks and sediments

The mantle and crust of the Earth largely consist of rocks. You must not think of a rock as a lump of material of foot-crushing size: rocks come in all sizes from dust grains to thick sheets which encircle the entire Earth. In the most general terms a *rock* is a solid body made up of naturally occurring chemical compounds, which in this context are usually called *minerals*. A rock may consist of just one type of mineral, or it may be a mixture of many different types aggregated in various sized pieces. If it lies at no great depth it may also contain holes, which may contain various gases or liquids.

There are three major types of rock: igneous, sedimentary and metamorphic.

An *igneous rock* is one which has been formed by the solidification of *magma*, which consists of liquid rock mixed with solid rock fragments. *Lava* is magma that has reached the surface of the Earth and flows from volcanoes and from other sources. The igneous rocks which result from the freezing of lava are called *extrusive* igneous rocks, because the magma has been extruded.

However, most igneous rocks regardless of whether they are exposed at the Earth's surface *today* solidified *beneath* the surface and are called *intrusive* igneous rocks, because they have intruded between or under pre-existing rocks.

The crust contains many different types of igneous rock, with silicates as the dominant compounds. However, in general terms the continental igneous rocks compared to the oceanic rocks contain comparatively large amounts of rocks dominated by silicon and oxygen and are also comparatively rich in aluminium, calcium, sodium and potassium. A familiar example is *granite*. By contrast, the oceanic igneous rocks contain slightly less silicon and oxygen, much less potassium, and rather more iron and magnesium. A widespread example is *basalt*.

The continental rocks have densities in crustal conditions ranging from about 2600 kg/m^3 to about 2900 kg/m^3 whereas the oceanic crustal rocks have densities ranging from about 2800 kg/m^3 to about 3200 kg/m^3. A few samples of upper mantle rocks have been obtained and these largely consist of silicates rich in iron and magnesium which have yet higher densities. There is also a general tendency for the range of melting temperatures to be higher in upper mantle rocks than in oceanic crustal rocks, and higher in oceanic crustal rocks than in continental crustal rocks.

Sedimentary rocks are produced at the Earth's surface from particles eroded from pre-existing rocks. Such particles can be transported over considerable distances, deposited, and subjected to various physical and chemical processes to yield sedimentary rocks. Much of the erosion of pre-existing rocks is due to the oceans and the atmosphere. Their erosive action is called *weathering*. There is a *physical* aspect to weathering, for example, the erosion of a rock by the wind and by frost, and also a *chemical* aspect, notably the chemical changes wrought by liquid water in which carbon dioxide is dissolved, which reacts chemically with the rock and in the process breaks it up. The oceans and the atmosphere are also prominent in the transport and deposition mechanisms of the eroded particles, and during this time further chemical changes can occur. Accumulations of deposited particles are called *sediments*. Such deposits can be turned into a sedimentary rock by chemical bonding of the particles, which is facilitated by *modest* pressures. For the sake of brevity I shall use the term *sedimentary deposit* to cover sediments and sedimentary rocks.

Under *high* pressures, and particularly if the temperature is raised, profound changes can be wrought in sedimentary rocks and also in igneous rocks, thus producing *metamorphic rocks*.

It has been estimated that about 95% of the volume of the Earth's crust consists of igneous rocks or of igneous rocks that have been changed into metamorphic rocks. However, most of these rocks are hidden by sedimentary deposits which cover the oceanic crust to an average depth of about 500 metres and which cover the continental crust to a somewhat greater depth. The most familiar covering of dry land is *soil*, which constitutes a thin veneer. Soil is a special form of sedimentary deposit because it contains living organisms and their byproducts.

In the chemical processes which help produce sedimentary rock volatiles can become chemically incorporated in the material. For example, in *clays*, which are a particular type of fine-grained sediment, there are molecules which are hydrated and which also contain the hydroxyl group (Table 2.2), and in the corresponding sedimentary rock, *claystone*, this is also the case. Another example is *limestone*, which consists largely of carbonates, and the carbonate group can incorporate carbon dioxide which is another volatile. The volatiles in these minerals can be released by subsequent metamorphic processes.

Ice conforms to the definition of a rock because it is a solid which consists of a mineral (water). However, ice is far more volatile than other rocks and it is therefore conventional to place it in a separate category. This category, which consists of volatile materials in their solid phase, I shall call *ices*, and water ice is now just one of many ices. No other significant ices can exist for long at the pressures and temperatures of the Earth's surface, but elsewhere in the Solar System other ices can exist, notably carbon dioxide ice (CO_2), ammonia ice (NH_3), and methane ice (CH_4). Regardless of whether they are in the solid phase I shall group the compounds H_2O, CO_2, NH_3, CH_4 and their mixtures in the category *icy materials*. Similarly, compounds which typify rocks are grouped in the category *rocky materials*, which may or may not include metallic iron, depending on the circumstances.

Another useful category is *carbonaceous materials*. These contain large amounts of carbon compounds, most of which are moderately volatile.

Small solid particles of icy, rocky and carbonaceous materials will be called *dust*, or *grains*, though some such particles will be rather soft.

2.1.10 Isostatic equilibrium and the asthenosphere

Gravitational studies have helped to reveal the architecture of the outer few hundred kilometres of the Earth as illustrated in Figure 2.9. But they have also revealed that these regions behave in the manner illustrated by Figure 2.10.

In this Figure the treacle crudely represents the asthenosphere, and the gelatin crudely represents the lithosphere. When the gelatin is slid into the treacle it gradually takes up the form shown in Figure 2.10 (b). It floats because it is less dense than the treacle, and it also bends until above each point on a horizontal surface placed at any depth which lies at or below the deepest part of the gelatin there is the same amount of mass. The highest level to which such a surface can be raised is called the *depth of compensation*.

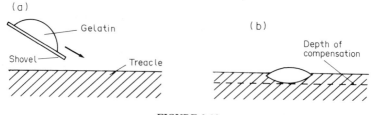

FIGURE 2.10
An example of isostatic equilibrium.

In Figure 2.10 (b) the depth of compensation is at the deepest point of the gelatin.

The equilibrium achieved in Figure 2.10 (b), in which there are equal masses above any point on a sufficiently deep horizontal surface, is called *isostatic equilibrium.* Isostatic equilibrium corresponds to the least energy of a system in which the upper layer is not appreciably plastic (section 2.1.2) but can bend, and in which the layer beneath is far more plastic and has greater density. If the upper layer is raised or lowered then gravitational forces are no longer balanced by molecular repulsion and a net force exists which restores the isostatic equilibrium.

Measurements of the force of gravity across the gelatin in Figure 2.10 (b) could, in principle, reveal whether isostatic equilibrium existed. In the case of the Earth such measurements have revealed that over most of the Earth's surface conditions are very close to isostatic equilibriuim. This reveals the presence of the asthenosphere as a plastic region, and also that the lithosphere is sufficiently flexible for isostatic equilibrium to be very nearly attained.

In Figure 2.9 you can see that beneath the continental crust the land surface approximately mirrors the Moho. This is similar to the gelatin model in that the crust varies in thickness from place to place but not appreciably in composition. A different case is posed by the change in composition from continental crust to the denser oceanic crust. Were oceanic crust as thick as continental crust then isostatic equilibrium would require the top of the oceanic crust to lie slightly deeper than the top of the continental crust. Such a slight height difference is increased to the observed value because the oceanic crust is thinner and because of the mass of the oceans above the oceanic crust which also has to be taken into account in isostatic equilibrium.

Remember that the asthenosphere is *solid* and that the lithosphere does not bob on it like a cork on water, nor for that matter like gelatin on treacle. If the lithosphere is subjected to a comparatively sudden alteration, such as the disappearance of an ice field in no more than a few hundred years, then the lithosphere will be temporarily out of isostatic equilibrium and it will take *thousands* of years to re-establish such equilibrium. An example of this is Scandinavia which is today rising at an observable rate. Estimates of the rate of rise, and of similarly caused rises elsewhere, have led to estimates of the plasticity of the asthenosphere. It seems that though the mantle is particularly plastic between depths of about 100 and 175 km it is far more plastic than the lithosphere right down to near the core and therefore the whole mantle below the lithosphere should be regarded as the asthenosphere. There is also some other geological evidence for such a thick plastic layer.

2.1.11 Plate Tectonics

Most of the large-scale features of the lithosphere and the processes which mould it can be explained by the theory of *Plate Tectonics*. This theory had a protracted birth, extending over 50 years from the first seeds sown in 1915 by the German scientist Alfred Wegener (1880–1930) to the full flowering and wide acceptance of the theory in the 1960s.

The essential features of the theory are as follows.

The lithosphere of the Earth is divided into six large *plates* plus a few smaller plates as shown in Figure 2.11. Each plate is created from the mantle by partial melting of the mantle at ridges which snake across the ocean floors. The plates slowly spread away from these ridges and where two plates meet the one dives beneath the other forming a trench in the ocean floor the diving plate being reincorporated into the mantle and hence losing its identity. At other places the plates slide past each other. An example of each of these cases is shown in Figure 2.11. Motion of the plates is facilitated by the plasticity of

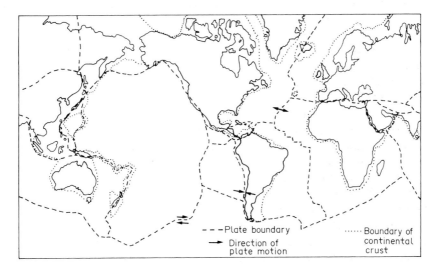

FIGURE 2.11
The lithospheric plates of the Earth.

the asthenosphere, and motion of the asthenosphere is an integral part of the motion of the plates, as outlined later.

The theory clearly gets part of its name from the central role played by these plates. *Tectonics* is used more widely to mean any large-scale deformation of rocks, and this in turn is a specific kind of *geological activity* which is the name for any kind of surface reworking driven from beneath the surface. *Volcanic activity* is another example of geological activity which is specifically related to the production of lava, ash and gases by volcanoes and by other sources.

Oceanic crust, according to the theory, is produced at ocean ridges as an integral part of the plate as shown in Figure 2.9 (a). The crust bears a high proportion of the more readily melted mantle materials and therefore the mantle part of the plate must be rather depleted in such materials. The mantle-part materials must also be denser than the oceanic crust otherwise they would have gravitationally separated upwards when the oceanic crust was molten. The mantle part of the plate probably consists largely of a silicate rock called *peridotite*. The lithospheric plate thus consists largely of oceanic crust and peridotite. Therefore the mantle beneath the plate consists of materials which can ultimately yield oceanic crust plus peridotite when this mantle material is partially melted.

Continental crust is a product of a further partial melting, this time at sites where one plate dives beneath another. Such sites are called *subduction zones* and an example is shown in Figure 2.9 (a). The source of heat for this partial melting is probably the result of friction between the two plates and the conduction of heat into the plate from the surrounding mantle. This heat melts some of the materials with the lower melting temperature ranges in the oceanic crust and in the mantle part of the *upper* plate. The amount of the melt and its composition are influenced by the presence of water in various forms particularly the hydroxyl form. The melted materials in the diving plate rise to form the continental crust. The remainder of the diving plate loses its identity in the mantle.

The comparatively low density of continental crust, lower even than that of oceanic crust, means that it is not readily taken down into the mantle. Thus, if two plates are pushing against each other, and one has continental crust at the boundary, then, as in Figure 2.9 (a), it is the *other* plate that will do the diving: the lower density continental crust acts as a "buoyancy device". Sedimentary deposits on top of oceanic crust also escape subduction, and collect in the subduction region, as shown in Figure 2.9 (a).

Continental crust tends to thicken above this subduction zone by a predominantly *vertical* addition from the diving plate. This can build large mountain ranges, and such regions can be further raised by uplift of the over-riding plate.

If the diving plate *also* carries continental crust then ultimately it will meet the continental crust on the over-riding plate. This will uplift any sedimentary deposits and bits of oceanic crust trapped between the two continental crusts. This uplift, plus any crumpling of the continental crusts, raises mountain ranges which do not border oceans. In this case the continental crusts tend to thicken as a result of the compression of the two plates, and this can cause thickening to extend some way from where the continents collided.

The thickening of the continental crust in all of these various regions displaces the oceanic crust from beneath it, so that today very little continental crust is underlain by oceanic crust.

It has been estimated that erosion is able to convert almost all of the present continental crust above sea level into sedimentary deposits on the ocean floors in only a few Ma (1Ma $\equiv 10^6$ years). Clearly the generation of new continental crust and its thickening is able to overcome this tendency.

The thickening of the continental crust also produces the observed greater thickness of continental crust than of oceanic crust (Figure 2.9 (a)), and isostatic equilibrium ensures that this continental crust extends, on average, to higher altitudes than oceanic crust. This, plus the continuous generation spreading and destruction of oceanic crust readily explains the twin-peaked hypsometric distribution of the Earth in Figure 2.9 (b).

It is clear that according to the theory there is a tendency for the amount of continental crust to grow with time and therefore that continental crust stretches back to the beginning of Plate Tectonics. By contrast oceanic crust, which is continuously being created and destroyed, is comparatively youthful. Powerful support for the theory comes from the measured ages of various crustal rocks, which show that oceanic crust near oceanic ridges is less than 1 Ma old whereas near subduction zones this crust is up to about 200 Ma old and that most continental crust is far older. These ages have been established by radiometric dating, which will be outlined in section 2.1.15.

Figure 2.12 shows what the Earth would look like were one to remove the oceans, the atmosphere and the *biosphere* (= living things and their surface remains). The predominance of separate altitudes for oceanic crust and continental crust is apparent. It is also clear from comparison with Figure 2.11 that departures from these two levels are concentrated on the whole along plate boundaries, in particular there are ocean ridges, ocean trenches and mountain ranges. Ocean trenches and continental mountain ranges border subduction zones, such as the Andes and the associated Peru-Chile trench. Bordering other subduction zones, at which there is less continental crust, there are mountain ranges called island arcs: the Aleutians are an example. The mid-Atlantic ridge is an example of an ocean ridge at which two plates are being created. The Himalayas are an example of mountain ranges raised when two continents collided.

The collision of continents is one of the main ways in which plate motion

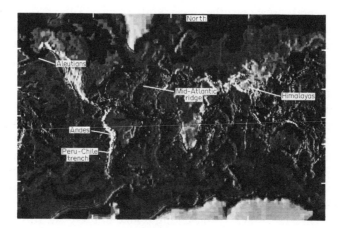

FIGURE 2.12
The Earth if it were devoid of atmosphere, oceans and biosphere.

can be changed. Such a collision will ultimately halt the motion of the one plate relative to another, and this can trigger new plate motion elsewhere, with the opening of a new oceanic ridge.

It is clear that plate motion, and the various changes in it that occur, carry continental crust to and fro. Some pieces are joined, others are broken up and thus the familiar map of the world is transitory. The relative motion between plates has been measured. A fairly typical value is 100 mm/yr which does not amount to much in a human lifetime but is rapid on the scale of the Earth's lifetime.

The explanatory power of Plate Tectonics is vast. But we must now turn to the question of what drives the plates. At the heart of the mechanism lies thermal convection.

● *2.1.12 Thermal convection*

Imagine a deep pan of liquid being heated from below. The layer at the bottom rises in temperature and therefore expands, and its density falls below that of the liquid above it. The gravitational energy of the system would be lower if the less dense liquid were on the top. However, the bottom layer cannot rise all in one piece and it is therefore possible for the liquid to acquire a temperature gradient across it with the heat being transported upwards only at a molecular level by means of the random motions of the molecules. This mode of heat transport is called *thermal conduction* or conduction for short. If the rate of heat input at the base is increased then the temperature gradient increases and the density disparity becomes even more marked. It can be shown that when the temperature gradient exceeds a value called the *adiabatic gradient* then the less dense layers below break up and rise through the denser layers above, which are displaced downwards. The mode of break up and the resulting pattern of circulation depends on the properties of the liquid and on the rate of input of heat, and the pattern can be extremely complicated.

This motion, driven by heat, is called *thermal convection* or convection for short. It supplements heat transport by conduction because the upward moving liquid loses heat as it approaches the upper surface. Once convection is established it usually dominates conduction as an upward heat-transport mechanism.

If the rate of heat input is further increased above the minimum required to establish the adiabatic gradient the convection becomes more vigorous and the rate of heat transport increases in such a way that the actual temperature gradient never greatly exceeds the adiabatic value. Thus, provided that the rate of heat input is sufficient to establish an adiabatic gradient the actual gradient will never be very much more than the adiabatic value. The value of the adiabatic gradient depends on the properties of the liquid and on the strength of the gravitational field across the liquid.

Thermal convection can also be established by cooling the top of a liquid rather than heating the bottom. Such convection will be short-lived if the liquid freezes, but this can be prevented by heat sources within the liquid. Moreover, these heat sources need not be at the bottom but could be distributed throughout the liquid.

2.1.13 Driving the plates

It is widely believed that the Earth's asthenosphere is in a state of thermal convection, and that this is intimately related to the forces driving the plates.

Convection in a solid material can only occur if it is sufficiently plastic, and even then it is so slow that it can only be of importance on a time scale in excess of thousands of years. Evidence that the asthenosphere is sufficiently plastic to permit such convection was outlined in section 2.1.10. Furthermore there is observational evidence for the slight horizontal variations in the mantle that would be associated with convection, seismic evidence being the most convincing.

The distribution of heat sources which drive this convection is not very well known and their nature is not certain: these problems will be taken up in section 2.1.16. However, it seems unlikely that the mantle is predominantly heated in its lower regions. It is more likely that it is heated throughout much of its volume and that cooling at its upper surface is an important means of driving the convection.

There are several forces that can exist between the plates and the convective motions. It is possible that the main force which sustains plate motion is provided by the lithospheric plate which dives at a subduction zone. That part of a plate which sits on top of the asthenosphere is slightly less dense than the asthenosphere. But when it has descended some way down the subduction zone it can be shown that it will be slightly more dense than the surrounding material. This is because its descent is sufficiently rapid that it fails to rise to the temperature of the surrounding medium, but the details need not concern us. It thus tends to sink, and pulls the rest of the plate behind it.

However, though there is no shortage of ideas about plate driving forces, and though it seems very likely that these are intimately related to mantle convection, the details as yet are far from clear.

● *2.1.14 Heat sources for a planet's interior*

A planet receives an initial energy supply from its accretion. Material added to a growing planet possesses considerable energy of motion just before impact, and this energy is lost largely in the form of heat which raises the temperature of the growing planet's interior. The energy of motion relative to the planet's surface was acquired largely from gravitational forces and therefore the *heat of accretion* represents a transfer of energy from gravity to heat.

One of the most important determinants of the degree of accretional heating is the final mass of the planet. It can be shown that for almost any feasible way in which a planet could accrete, a planet which grows to only a small mass will not, at the end of accretion, possess as much heat of accretion per unit mass as a considerably more massive planet. Thus the interior temperatures of high mass planets are expected to be initially higher than those of low mass planets. For example, Jupiter is expected to have become *much* hotter than Mars.

It is possible that the tail end of accretion can be a protracted process, and this can be an important means of heating the surface regions in the early part of a planet's lifetime.

Another source of energy is also gravitational and arises from the differentiation of a planet which initially has denser substances further from the centre than the less dense substances. If the interior is sufficiently yielding then the denser substances will move downwards, displacing the less dense substances upwards. The denser substances come to rest nearer the centre which means they will have initially acquired energy of motion from gravity and then lost it in the form of heat, and the same is true of the substances which moved upwards. This is *heat of differentiation*.

A completely different source of heat is provided by *radioactive isotopes*. An atom of a radioactive isotope, *without* external provocation, "self-destructs" to form an atom of an isotope of another element. This is called *radioactive decay*. In such a decay various atomic particles are emitted. These particles, and any motion of the resulting isotope, can be slowed down by the surrounding material and therefore radioactive decay acts as a source of heat, *radioacive heat*.

There are also some sources of heat that are usually small compared with those outlined above. For example, there is heat from solar radiation at a planet's surface. Also, if there is a satellite near the planet then there may be significant tidal heating (section 6.1.1). Chemical reactions, phase changes and thermal contraction can also provide heat.

Not all the heat released by all these various processes necessarily goes to raise the interior temperatures of a planet. For example, the three last-named processes, namely, chemical reactions, phase changes and thermal expansion (rather than contraction), can *absorb* heat from other processes without the associated material rising in temperature.

The accretional heat source is concentrated near the beginning of a planet's lifetime, and if there were no other appreciable heat sources then it would not be long before the interior temperatures would start to decline. The heat from *short-lived* isotopes (section 2.1.15) is also concentrated into a short time. Radioactive heat from *long-lived* isotopes (section 2.1.15) is also generated at the greatest rate near the beginning but the decline in heat generation is very slow.

A declining heat source will yield a temperature rise that continues after the heat source itself was at maximum. With zero heat losses the temperature would rise, reaching a constant value when the heat source is exhausted. But heat *is* lost (to space), and therefore a declining heat source can cause a rise in internal temperatures until the rate of loss of heat matches the rate of generation. In the case of long-lived isotopes this can lead to an increase in internal temperatures to some maximum values long after the planet was formed. Moreover, if at the maximum values thermal convection is occurring in the planet then the maximum temperatures can be sustained for some time, the rate of convection, but not the temperature, falling as the rate of heat generation continues to decline. A rough analogy is the heating of a kettle of water on a camping stove the supply of gas for which is slowly running out. As the gas flame dies the water temperature continues to rise provided that the rate of heat input exceeds the rate of loss. If the kettle comes to the boil it can be sustained there by the last vestiges of the gas.

A time scale for these events can be established for a planet by means of radiometric dating.

● *2.1.15 Radiometric dating*

Until about the turn of the century distant events in Earth history could only be placed in a time-order. A *time-order* of events shows the order in which the events happened but not how long ago each event happened. Then came radiometric dating.

Radiometric dating is based on the decay of radioactive isotopes (section 2.1.14). You will recall (section 2.1.1) that each element can exist in the form of several isotopes. In the case of uranium *all* known isotopes are radioactive whereas for most of the elements only some of the isotopes are, the rest being *stable*.

Of particular interest to radiometric dating is the element potassium (K). Potassium is a fairly common element in the Earth's crustal rocks and in the oceans. Of these potassium atoms about 93.08% have mass-number 39 (^{39}K) and about 6.91% have mass-number 41 (^{41}K). Neither of these isotopes is radioactive. But ^{40}K, which accounts for about 0.012% of the potassium atoms, decays to form ^{40}Ar, a stable isotope of argon. In the decay a proton in the nucleus of ^{40}K turns into a neutron by capturing one of the electrons surrounding the nucleus and emits a gamma ray in the process (section 2.2.1) which has negligible mass. Thus, the argon nucleus has the same mass-number as the potassium nucleus but the atomic-number is one less. The neutral form of ^{40}Ar thus has one electron fewer than ^{40}K and this leads to considerable chemical differences between them.

Laboratory studies long ago showed that the decay of a particular atom occurs in a tiny fraction of a second but that not all the atoms of a particular isotope decay at the same time. Instead there is an average lifetime with individual atoms decaying at random. This can be shown to correspond to an *exponential decline* in the number of atoms of a radioactive isotope, as illustrated in Figure 2.13 for ^{40}K. You can see that after about 1200 Ma the number of ^{40}K atoms remaining is half that at the start. The exact figure is 1193 Ma. This is called the *half-life* of ^{40}K and it is independent of external factors such as pressure and temperature. It is a feature of an exponential decline that an interval equal to the half-life can be placed anywhere on the time axis and the number of atoms at the end of the interval is half that at the

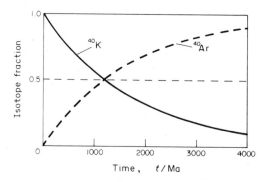

FIGURE 2.13
The exponential decline of ^{40}K through radioactive decay.

start: the starting time makes no difference — try it. As the ^{40}K declines the ^{40}Ar grows, and because ^{40}Ar is stable the growth curve mirrors the ^{40}K decay curve as shown in Figure 2.13.

When a rock melts, any ^{40}Ar, being a gas in crustal and mantle conditions, bubbles upwards and escapes. Therefore, when the rock solidifies it is free of ^{40}Ar. In Figure 2.13 this corresponds to $t = 0$. If there is any potassium present then the ^{40}K will decay and the ^{40}Ar will grow. If the ^{40}Ar does not escape and we observe today a certain ratio of ^{40}Ar to ^{40}K then the time since $t = 0$ can be determined. For example, if the ratio of ^{40}Ar to ^{40}K is about 3:1 then you can see from Figure 2.13 that this corresponds to about $t = 2400$ Ma. Therefore this rock solidified about 2400 Ma ago. Condensation from the vapour and strong heating without melting could also be dated. This is *potassium-argon dating*, and in its basic features it typifies *radiometric dating*.

Similar kinds of events can also be dated by means of other isotopes. Of particular importance are ^{87}Rb (Rb = rubidium) which decays to ^{87}Sr (Sr = strontium), and ^{238}U and ^{235}U (U = uranium) which decay through a complex sequence of transformations to yield various stable isotopes of lead (Pb). None of these stable end products are gases and therefore any separation of the starting isotope from its end product to define $t = 0$ rests on their different chemical properties.

The best isotope to use to date an event depends on the particular circumstances. The isotopes ^{40}K, ^{87}Rb, ^{238}U and ^{235}U between them cover most circumstances. Their half-lives differ from one to another and overall they can date events from about 500 000 Ma ago to about 1 Ma ago. Their half-lives are all fairly long, and therefore they are called *long-lived isotopes*. In order to date more recent events *short-lived isotopes* with shorter half-lives can be used, and use can also be made of non-radiometric methods. In all cases precautions have to be taken to ensure that an accurate date is obtained. For example, after some event which defines $t = 0$ the preferential escape of the end isotope can make the apparent date of the event seem *less* far back in time than it really is.

^{40}K, and the other radioactive isotopes listed above and in section 2.1.16, have never been produced in the Earth in appreciable quantities and therefore the Earth's supply must always have been running down. Most scientists believe that the PFM obtained the initial supply from exploding stars called *supernovae*.

The absence of any replenishment also means that radioactive heat from all the isotopes of importance to radioactive heating has been in decline since the beginning. Clearly it is the half-life which determines the span of time over which radioactive heat from an isotope is appreciable.

A rather different type of event has been dated with the aid of radioactive isotopes, and this is the *age of the Earth*. The method uses the relative abundances in the Earth of certain lead isotopes, in particular ^{204}Pb, which is *not* produced by radioactive decay, and ^{206}Pb and ^{207}Pb, of which at least some of the Earth's supply has come from the radioactive decay of ^{238}U and ^{235}U respectively. The argument is different from that of the radiometric dating outlined in the preceding section, and the type of event dated is also different. The argument is intricate and I shall not go into details, but on the basis of certain plausible assumptions the argument leads to the conclusion that the uranium which the Earth received was probably incorporated into condensed bodies about 4600 Ma ago. These bodies would have been among those from which the Earth accreted, and most theories of accretion suggest that this did not take very long, making the Earth about 4600 Ma old.

Further support for this age comes from meteorites. The radiometric ages of various materials in meteorites stretch back to about 4600 Ma, and this is also the case for samples of lunar materials.

2.1.16 *Heat sources and losses, and the evolution of the Earth*

It is not known how hot the Earth became when it formed, but it is highly likely that the Earth is sufficiently massive for nearly all of the interior to have been at least warm. Nor is it known whether the Earth accreted homogeneously, but if it did then subsequent differentiation would have provided a substantial additional source of heat. Growth of the inner core could be a minor source of heat of differentiation today.

Radioactive heating is an embarrassingly large source. In the present crustal rocks most of the radioactive heating comes from the decay of the isotopes ^{238}U, ^{235}U, ^{232}Th (Th = thorium) and ^{40}K. If the whole Earth today contained these four isotopes in the same abundances as in the crustal rocks then the interior temperatures would be far higher than any external observations allow. Therefore, some or all of these isotopes must be less abundant in the whole Earth than in the crustal rocks. If this was not originally the case then chemical differentiation is required to give a measure of upward segregation, and there are plausible schemes for this. Alternatively, these isotopes could have been more abundant in the materials which accreted later than in those which accreted earlier. In the distant past there may also have been short-lived radioactive isotopes such as ^{26}Al (Al = aluminium).

Substantial tidal heating could have been caused by the Moon in the distant past if it was then a good deal closer to the Earth than it is today: more on this in Chapter 6. Chemical reactions could only have been a significant source of heat if the Earth accreted homogeneously in such a way that there was initially an intimate mix of iron and oxygen-rich materials. Solar radiation has always been a negligible source of energy for the Earth's interior.

The loss of energy from the interior of the Earth is almost entirely in the form of heat transferred from the interior to the surface. This raises the surface temperature very slightly above what otherwise would have been the case and this increases the rate at which the surface loses energy upwards, ultimately by radiation to space. It is important to realize that the surface temperature of the Earth is only *very slightly* raised by this flow of heat from the interior. By far the major input of energy to the Earth's surface is solar radiation, which deposits energy at the surface at a globally averaged rate about 5000 times that of the flow of heat from the interior. However, because there is no convection in the *crust*, and because its thermal conductivity is not high, most of the heat from solar radiation is confined to the outer few *metres* of the crust where its effect is to raise the surface temperature to about 288 K from the temperature fairly close to absolute-zero that would otherwise prevail. The temperature rises no further because at this temperature the rate of energy input to the surface equals the rate of loss.

The rate of loss of energy from the surface, including the small contribution from the interior, is proportional to the surface area of a planet, whereas the amount of interior energy is roughly proportional to the mass of the planet, and the greater the mass the smaller the ratio of surface area to mass. Therefore, the more massive the planet the smaller the rate of energy loss per unit mass and the longer it can retain internal heat.

Because the loss of heat from the interior of a planet occurs largely from its surface the internal temperatures evolve so that ultimately there is a general increase in temperature towards the centre.

This feature is apparent in Figure 2.3 where the temperatures in the Earth increase fairly smoothly from about 288 K at the surface to several thousand K at the centre. These model temperatures correspond to conduction across the lithosphere, a slightly greater than adiabatic gradient in the asthenosphere which extends near to the core, an outer core temperature just above its freezing-point and an inner core temperature just below its freezing-point. There are plausible arguments which justify the choice of such temperatures in the core, and such temperatures clearly depend on the composition of the core, different models with different core compositions having core temperatures which differ by up to 1000 K or so from those in Figure 2.3. The near adiabatic gradient in the asthenosphere also depends on its composition, though here the range of possibilities is a good deal less.

The temperature drop across the lithosphere is very large, far in excess of the adiabatic gradient. However, plasticity declines very rapidly with falling temperature and therefore the lithosphere is not plastic enough to permit convection at an appreciable rate. The lithosphere can thus sustain the large temperature drop, which corresponds to heat transfer by the relatively inefficient process of conduction. However, the lithosphere is an integral part of mantle convection its motion being parallel to the Earth's surface. It is therefore *not* a static blanket in which the Earth is wrapped.

The temperatures in Figure 2.3 are consistent with the external observations. You can probably match them qualitatively to evidence from seismic observations, the magnetic field, isostasy, the requirements of Plate Tectonics, the possible heat sources. Direct observations of temperatures are confined to the outer few kilometres of the crust, and show that on average the temperature gradient is about 25 K/km near the surface, the gradient increasing with depth. The existence of the low velocity zone (section 2.1.6) and of magma indicates that in parts of the upper mantle the temperatures are high enough to partially melt some of the materials there.

The temperatures in Figure 2.3 are the outcome over the history of the Earth of all the various energy sources that have fed energy into the interior and of the various rates at which this energy has been lost. It is not possible to deduce how the internal temperatures have evolved, nor correspondingly how the interior of the Earth itself has evolved, but here are some broad guidelines.

The accretion of the Earth was largely complete by about 4600 Ma ago, though it is not known whether it accreted homogeneously or heterogeneously. An argument against homogeneous accretion has been raised, namely, that were the core and mantle to be mixed today then some of the iron would combine with oxygen drawn from oxygen-rich compounds in the mantle. What prevented this happening if they were originally mixed? However, it is not certain that much iron would combine with oxygen, and in any case there are ways of circumventing the possible problem, for example if the iron-rich and oxygen-rich bodies were at least several kilometres in radius.

It seems likely that the heat of accretion, perhaps with the assistance of short-lived radioactive isotopes such as ^{26}Al, was sufficient for the materials which now make up the core to melt throughout much of the Earth. Thus, regardless of whether the Earth accreted homogeneously, or heterogeneously with its core accreting first, it seems likely that within 100 Ma or so of the Earth's formation the core was in existence and was largely liquid. Rocks as old as 3700 Ma show evidence of the presence of a magnetic field when they solidified, which indicates that the internal magnetic dipole field might have been in existence, though it is only in rocks up to 2600 Ma old that there is *clear* and continuing evidence for this. However, these dates certainly do *not* rule out the possibility of much earlier liquid core formation.

Because of their higher melting temperatures it seems less likely that the mantle materials were largely molten. In any case, had they become molten then convection in a *liquid* mantle would be so rapid that the mantle would soon lose enough heat to solidify.

Most scientists believe that the Earth has contained sufficient long-lived radioactive isotopes to sustain convection in a *solid* mantle from the earliest times to the present. By contrast, the *surface* of the Earth, because of the loss of energy to space, has always been cold enough to be rigid, except perhaps regionally as a result of any protracted tailing off of accretion or of any subsequent heavy bombardment. However, in these earliest times this rigid outer surface does not anywhere correspond to the continental crust. The oldest known crustal rocks are certain metamorphic rocks in Greenland. Radiometric dating has revealed that these rocks were last strongly heated about 3800 Ma ago, and other evidence (I shall not present the details) indicates that this was when these rocks were derived from the mantle. Thus, the processes which give rise to the continental crust do not seem to have become established until about 3800 Ma ago. Had they occurred earlier then there is a good chance that such crust would have survived, because its comparatively low density would have prevented it being readily taken down into the interior. By contrast the rigid surfaces that formed before 3800 Ma ago seem to have been removed to the interior, though any melting through heavy bombardment would have helped to erase them.

In correlating the ages of continental rocks with their nature and distribution it has been established that a global system of large and mobile lithospheric plates of the sort that exist today was probably not established until about 2200 Ma ago. Before that time there certainly seems to have been lots of action, but of a rather different nature. There is also evidence for less profound changes in lithospheric activity at other times. These large and small changes in lithospheric activity reflect changes in the pattern of mantle convection. Such changes are expected in response to the fall in the rate of generation of heat from all sources, though remember from section 2.1.14 that this does *not* mean that *temperatures* have necessarily been declining: indeed convection in much of the interior tends to hold temperatures near the adiabatic gradients, and the changes in the pattern of mantle convection are mainly in response to the fall in the rate at which the mantle has to transfer heat to the conducting lithosphere. Thus, the temperatures in the Earth might have been much like those in Figure 2.3 for thousands of Ma, but the rate of convection might have been much higher in the past.

At some stage the interior temperatures must begin to fall appreciably. This will be most marked in and just below the lithosphere and will lead to a thickening of the lithosphere as the cooled upper asthenosphere loses its plasticity. Ultimately this will lead to a cessation of geological activity and to a consequent loss of dry land through erosion. The core will solidify and the dipole field vanish. Some thickening of the lithosphere has probably already started, but the geologically moribound Earth probably lies thousands of Ma in the future.

2.2 The Earth's atmosphere and surface volatiles

The Earth's atmosphere is one of several repositories of volatiles. Amongst the other repositories are a group which I shall call the *surface repositories*.

This group consists of sedimentary deposits, metamorphosed sedimentary rocks, the biosphere, and the oceans plus other bodies of water including ice. The volatiles in the surface repositories I shall call *surface volatiles*. The remaining repositories are the igneous crustal rocks (and their metamorphosed products) and the interior.

Figure 2.14 shows the present distribution of the more abundant volatiles in the atmosphere and in the surface repositories. The *volume* of each cube is proportional to the *number* of molecules of the kind given. You can gauge the absolute quantities from the corresponding mass of water in open liquid bodies, mainly the oceans, which is 1.5×10^{21} kg, or 0.025% of the mass of the Earth. Clearly, all these volatiles account for only a tiny fraction of the Earth's mass.

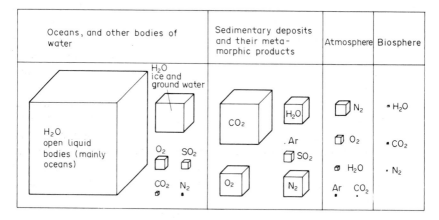

FIGURE 2.14
The distribution of the Earth's more abundant volatiles among various repositories.

Note that the molecular form given in Figure 2.14 corresponds to the common form of the volatile were it in the atmosphere. In the other repositories the molecular form can be different, as follows. In the oceans much of the CO_2 is in *bicarbonate* form, that is HCO_3, and in the rocks much of it is in the form of carbonates (CO_3 group). In the oceans and rocks much of the SO_2 is in sulphates (SO_4 group) and much of the nitrogen is present as nitrates (NO_3 group) and nitrites (NO_2) group. In the rocks much of the water is present in hydrated compounds (H_2O attached) or in hydroxyl form (OH). In arriving at the relative numbers of molecules in Figure 2.14 allowance must be made for these differences between the molecular form given in Figure 2.14 and the actual form. For example, the carbonate (CO_3) group is richer in oxygen than CO_2 by one atom. Thus, in the rock repository of CO_2 in Figure 2.14, for each CO_2 molecule that actually corresponds to a

CO_3 group there is one atom of oxygen added to the rock repository of O_2 and this accounts for much of the O_2.

The igneous crustal rocks and the interior have been excluded largely because it is extremely uncertain how much of the various volatiles they contain. However, it is certain that they contain small quantities of nearly all the volatiles shown in Figure 2.14, and though such volatiles account for only a tiny fraction of the Earth's mass they could add significant amounts to the volatiles in Figure 2.14.

The various volatile repositories are not isolated from each other. There is a continuous exchange between them of various volatiles at a variety of rates and in a variety of ways. The exchange rates are controlled by atmospheric, geological and biospherical activity. The role of the biosphere will be outlined in section 2.4 and that of geological activity in Chapter 5. There is also a loss to space from high in the atmosphere (section 2.2.2), particularly of low mass atoms and molecules such as hydrogen liberated from water. If, for a particular repository,the loss rate of some volatile is not equal to the gain rate then the amount of that volatile in that repository will be changing, and it will stop changing only when it is of such a size and location that the loss and gain rates are equal.

It is important not to confuse an element in a volatile form with the same element in non-volatile form. For example, oxygen is a prominent volatile in Figure 2.14 and yet there are *far* larger quantities of the element in the non-volatile silicate and oxide forms which account for much of the crust and the mantle. Indeed, oxygen is the most common element in the Earth. But though this oxygen can to some extent be exchanged with oxygen in volatile form the total amount in non-volatile form has hardly varied on a global scale. Nevertheless, any crustal igneous rocks which have been exposed on the Earth's surface and which initially contained less oxygen in non-volatile form than they could have done, would extract oxygen from one of the volatile repositories, and in this way act as a one-way drain on volatile oxygen. The quantity of volatile oxygen lost in this way is not known, but it could account for more oxygen than in all the O_2 repositories in Figure 2.14: more on this in section 2.4.

Much of the rest of section 2.2 is devoted to the Earth's atmosphere, but first you need to know something about electromagnetic waves.

● *2.2.1 Electromagnetic waves*

Light is a familiar example of an *electromagnetic wave*. As their name implies they consist of wave-like variations in magnetic and electric fields. Imagine that you are standing at P in Figure 2.15 and that you are holding a sensor of such fields. If the simplest kind of electromagnetic wave passes by then it would cause the signal from the sensor to vary with time as shown.

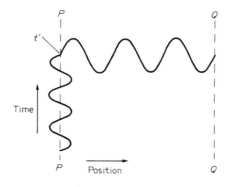

FIGURE 2.15
The variation with time and position of the output from a sensor of a simple (sinusoidal) wave.

The number of complete cycles per second is called the *frequency* of the wave. If you now imagine "freezing" time at t' in Figure 2.15 and moving through space from P to Q then a similar type of variation in the signal from the sensor will be obtained, but now it is a variation in space, and the cycle length is called the *wavelength*.

Electromagnetic waves of any frequency can travel through a vacuum. Such waves can also travel through matter, though a particular material will block certain frequency ranges, depending on the material. Waves which do penetrate matter suffer a *wavelength* change, the frequency remaining constant. Associated with such wavelength changes is a change in wave speed, in accord with the relationship speed = wavelength × frequency. In a vacuum the speed of electromagnetic waves is 2.998×10^8 m/s. This is called the *speed of light*.

Electromagnetic waves are usually classified according to their wavelengths in a vacuum. Such a classification is shown in Figure 2.16. Note that the wavelength scale is logarithmic, which enables the enormous range of wavelengths to be shown, from gamma rays, which have extremely short wavelengths, to radio waves, which have extremely long wavelengths. Light is the name of electromagnetic waves to which our eyes are sensitive, and you can see from Figure 2.16 that they occupy a very small range of wavelengths: we hardly "see" the Universe at all.

Electromagnetic waves are generated by a great variety of means. Of particular importance are *thermal sources*, that is, sources which emit because of the random molecular motions within them. All substances have such motions and therefore all substances are thermal sources and emit *thermal radiation*.The wavelength range and the amount of radiation depends strongly on the temperature of the source, the higher the temperature the greater the random molecular motions the greater the amount of radiation

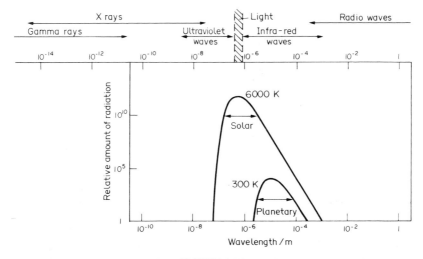

FIGURE 2.16

Classification of electromagnetic waves according to their wavelengths in a
vacuum, and the thermal wave spectrum of a body at 6000 K and another body at
300 K.

and the more it is shifted to shorter wavelengths. Figure 2.16 shows the
radiation from the surface of an opaque body at 6000 K, which is close to the
surface temperature of the Sun, and from a body of similar size at a surface
temperature of 300 K, which is typical of the surface of the Earth. I shall call
the solar wavelength range *solar wavelengths* and the terrestrial wavelength
range *planetary wavelengths*. The solar wavelength range will also be called
optical wavelengths because of the so-called optical techniques that are used in
instruments that operate over much of this range.

Any display of amount of radiation versus wavelength, such as those in
Figure 2.16, is called a *spectrum*.

I shall often abbreviate electromagnetic to *em*.

2.2.2 The atmosphere today

Figure 2.17 shows the vertical structure of the Earth's atmosphere. This is
an averaged picture: there will be small variations with time and space.

The amount of atmosphere is expressed in kilogrammes per square metre
of the Earth's surface, abbreviated kg/m². This is called the *column mass*. The
associated pressure depends on the Earth's gravitational field. The zero of
altitude in Figure 2.17 corresponds to sea-level, at which the average surface
pressure is 1.013 bar, which corresponds to a globally averaged column mass
of 10 360 kg/m². The mass of atmosphere above a given altitude must clearly
decrease with increasing altitude. This decrease is particularly rapid because

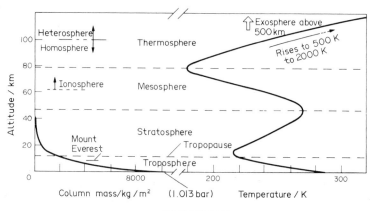

FIGURE 2.17
The vertical structure of the Earth's atmosphere.

gases are readily compressed and therefore as the altitude decreases the density will increase fairly rapidly. About 90% of the mass of the atmosphere lies below an altitude of about 16 km. Above an altitude of about 40 km the column mass of the remaining atmosphere is too small to be shown in Figure 2.17 but the atmosphere is still tailing off at far higher altitudes until it merges imperceptibly with interplanetary space.

You can see from Figure 2.17 that over certain altitude ranges the temperature *decreases* with increasing altitude and that over other ranges it *increases*. These different horizontal layers are given names: from the bottom up, troposphere, stratosphere, mesosphere and thermosphere, as shown in Figure 2.17.

In the *troposphere* the atmosphere is predominantly heated from below by the absorption of solar radiation at the Earth's surface which raises the temperature of the surface to a global average of about 288 K. This has resulted in thermal convection in the troposphere and to a corresponding temperature gradient close to the adiabatic value, as outlined in section 2.1.12. In the case of an atmosphere the adiabatic gradient is called the *adiabatic lapse-rate*. This is a very stable gradient in the troposphere: if the actual gradient were less than the adiabatic value then convection would cease, the lower layers would lose solar heat relatively slowly, and their temperatures would rise until the adiabatic lapse rate and convection were re-established; if the actual gradient were greater than the adiabatic value then the energy transfer by convection would be so rapid that the temperatures of the lower layers would fall until the adiabatic lapse rate were again re-established.

The value of the adiabatic lapse-rate depends on the strength of the gravitational field across the atmosphere and on the properties of the gaseous constituents, including those which can condense, notably water in the case of the Earth's atmosphere. In dry parts of the troposphere the value is about 10 K/km, whereas in damp parts it is about 5 K/km.

The condensation of water releases energy, and this is an additional and important means of transferring energy in the Earth's atmosphere. It is *not* the same as convection, but is an example of latent heat (section 3.4.2).

I want now to transfer attention to the top of the atmosphere, to the *thermosphere* (Figure 2.17). Here the lapse-rate is negative, that is, the temperature *increases* with altitude. In the upper thermosphere ultraviolet radiation at wavelengths less than about 10^{-7} m is absorbed because it ionizes some of the atoms there. Such radiation would ionize atoms lower down except that it is completely absorbed at these high altitudes. There is also some absorption of ultraviolet (uv) radiation at wavelengths between 10^{-7} m and 3×10^{-7} m through the disruption of molecules by em waves: this is called *photodissociation*. All of this uv absorption places energy directly into the upper reaches of the thermosphere and this is supplemented by energy of motion of particles emitted by the Sun, the *solar wind*, to be outlined in Chapter 9. The outcome is an atmospheric layer that is largely heated from above. In this case there can be no tendency to establish an adiabatic lapse rate. The atmosphere ends up with the observed negative lapse rate. There is no convection and therefore energy transfer is by thermal conduction and by radiation mainly at infrared wavelengths (Figure 2.16). Thermal conduction is a relatively inefficient process, and moreover the molecules in the thermosphere are largely of a type that are weak radiators. Thus, the upper thermosphere gets rather hot, as indicated in Figure 2.17.

Solar uv radiation and the solar wind are subject to modest variations and therefore the temperature of the thermosphere is also variable.

High within the thermosphere, above an altitude of about 500 km, the *exosphere* is encountered. In the exosphere the density of the atmosphere is so low that an atom or molecule rarely collides with another. Therefore, if an atom or a molecule is travelling away from the Earth sufficiently fast to escape then it very probably *will* escape. The requisite speed is determined by the gravitational field. Atoms or molecules entering the thermosphere from below usually have speeds determined by the temperature at the base of the thermosphere. The higher the temperature the higher the speeds and the greater the rate at which a particular type of molecule will escape. The more massive a molecule the slower its speed at a given temperature and the smaller the rate of escape. Escape determined by temperature is called *thermal escape*. Chemical reactions can accelerate participating molecules, and escape resulting from speeds increased in this way is called *chemical escape*. At present the rate of escape of atoms and molecules from the Earth's atmosphere is negligible except for atomic and molecular hydrogen, and helium, these being of low mass.

In the *mesosphere* and in the *stratosphere* most of the uv radiation which passes through the thermosphere is absorbed. This accounts for most of the solar radiation in the wavelength range 10^{-7} m to 3×10^{-7} m. The absorption is mainly by *ozone*. This is produced by the photodissociation of oxygen molecules (O_2). This yields oxygen atoms some of which combine with undissociated oxygen molecules to produce ozone (O_3). The initial photodissociation of O_2 absorbs some of the uv radiation, but absorption by ozone itself accounts for most. Were there no ozone then much of the uv radiation which it absorbs would reach the Earth's surface where it would prove lethal to many life forms including ourselves.

The temperature "bulge" in the stratosphere and mesosphere (Figure 2.17) is the combined result of an *increase* in atmospheric density with declining altitude with a consequent increase in the amount of O_2 available for conversion to ozone which absorbs the uv radiation, and a *decrease* with declining altitudes of the amount of uv radiation available for absorption.

Absorption of uv radiation in the atmosphere, and of solar wind particles high in the thermosphere, not only raises temperatures but also produces ionization. The level of ionization is significant in the upper mesosphere and in the thermosphere, and these regions constitute the *ionosphere*. The ionosphere is an example of a *plasma*: this is a neutral gaseous medium in which an appreciable fraction of the atoms and molecules are ionized, with a corresponding fraction of *free* electrons. The dominant ions in the Earth's ionosphere are O^+ and O_2^+. The ion content is variable, being particularly sensitive to solar uv radiation. Moreover, the ion content responds quickly to changes in uv radiation and therefore during the night, when solar uv radiation is absent, the ion content is noticeably lower. By contrast, temperatures do not respond so quickly and their day-night cycle is less marked. The ionosphere has been of practical importance because it can reflect radio-waves of certain wavelengths, thus facilitating long-distance communication.

The atmosphere can also be divided into a *homosphere* and a *heterosphere* (Figure 2.17). In the homosphere the relative abundances of the various gases are independent of altitude (except for those which can condense, particularly water). In the heterosphere molecular interactions are sufficiently slight that the different types of molecule separate to some extent, the more massive the molecule the more it is confined to lower altitudes.

2.2.3 Climate

The *climate* of a region is specified by the behaviour of its weather over several decades.

One characteristic of this behaviour is the seasonal variations, spring, summer, autumn and winter. This seasonal cycle arises from the annual variation in the amount of solar radiation that any region of the Earth's surface receives, solar radiation being by far the largest source of energy for the Earth's climatic system. This variation is largely the result of the axial inclination of the Earth, as illustrated in Figure 2.18 which shows the Earth's orbit to scale but with the diameter of the Earth in the orbit exaggerated about 1400 times. You can see that the region indicated by R in the enlargements of the Earth receive more solar radiation per unit surface area in June than in December. By comparison the variation of the distance of the Earth from the Sun arising from the (slight) eccentricity of the Earth's orbit is of minor importance. Roughly speaking, the seasonal variations increase with increasing latitude.

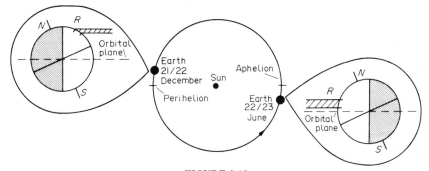

FIGURE 2.18
The Earth's seasons. Note that the enlargements of the Earth are viewed *edgewise* to the Earth's orbit.

The Earth's climatic system would be relatively simple if the climate of each latitude were independent of that at any other latitude. However, there is a considerable transfer of energy across latitudes from equatorial regions, which receive large amount of solar radiation, to polar regions, which receive small amounts. This transfer is accomplished by large-scale motions of the atmosphere and the oceans. These motions are "driven" by the equator to pole differences in solar radiation and they greatly reduce the climatic changes with latitude that would otherwise exist, in which the poles would be far colder and the equator rather warmer than they actually are. These large-scale motions are called *atmospheric circulations* and they are *extremely* complicated and varied. It is therefore *extremely* difficult to understand in any detail the determinants of the climate of any particular region of the Earth. It is rather easier to understand, in broad outline, the determinants of the globally and seasonally averaged climate.

One of the most important of these determinants is the rate at which solar radiation is absorbed by the atmosphere, oceans, and surface of the Earth because solar radiation is by far the largest source of energy for the Earth's climatic system. This rate clearly depends on the *luminosity* of the Sun, which is the rate at which it radiates energy to space, and on the distance of the Earth from the Sun, which determines the solar radiation available at the top of the atmosphere. But not all of this radiation is absorbed. A surface will scatter some of the incident solar radiation back to space. The ratio of the scattered to the incident solar radiation is called the *albedo* of the surface. The average value for a whole planet is called the *Bond albedo*, after George Phillips Bond (1825–1865) the American astronomer. The Bond albedo of the Earth is 0.33. In Figure 2.1 you can see that the albedos of the oceans and continents are low, whereas clouds have a high albedo and account for most of the solar radiation scattered back to space.

Most of the radiation which is not scattered back to space by a cloud emerges through its base, that is, the *absorption* in the cloud is small. The ratio of the radiation passing through a medium to the incident radiation is called the *transparency* of the medium. The complementary quantity is the *opacity* of the medium.

Clouds are an example of an *aerosol*, which is any suspension of solid particles or liquid droplets. The clouds in the Earth's atmosphere are collections of water droplets or water ice particles, and in many cases the number of particles or droplets per unit volume, that is, the *number density* is very high. Any aerosol of low number density is called a *haze*. However, even in clouds with a high number density of comparatively large particles or droplets the particles or droplets account for only a tiny fraction of the volume of the cloud, most of the volume consisting of the gaseous atmosphere between the particles or droplets. Other important aerosols in the Earth's atmosphere are dust raised from the surface, and particles produced by the action of solar radiation on gaseous sulphur compounds, notably on sulphur dioxide (SO_2). Most of the haze and clouds in the Earth's atmosphere is in the troposphere.

The Earth's atmosphere and surface are close to a state of energy balance in which the Earth radiates energy to space at very nearly the same rate at which the atmosphere and surface absorb energy from all sources. The rate of radiation to space can be expressed in terms of an average temperature, called the *effective temperature*. If, as in the case of the Earth, the largest source of energy for the atmosphere and surface is solar radiation then the effective temperature depends almost only on the solar luminosity, the solar distance, and the Bond albedo. In the case of the Earth the effective temperature is 253 K.

However, the actual temperature at any point in the atmosphere or at the surface can differ significantly from this effective value. In the case of the Earth the surface temperatures are significantly raised by the *greenhouse effect*, which gets its name from the initial but erroneous belief that it explained how greenhouse interiors become warm in the Sun. The name, however, sticks.

The greenhouse effect is the name given to the phenomenon whereby the temperature of the surface of a planet, and the atmosphere in contact with that surface, is raised because the atmosphere is more transparent to solar radiation than to radiation from the planet's surface. This is possible because the Sun is far hotter than a planet's surface and therefore as illustrated in Figure 2.16 there is a considerable separation in the wavelengths associated with the corresponding thermal radiation spectra. The rise in temperature occurs as follows. The solar radiation which reaches the lower atmosphere is mainly absorbed by the surface. The surface temperature rises until there is an energy balance at the surface between the absorbed solar radiation and the energy lost by the surface. Rocks are good insulators and therefore a negligible amount is lost into the interior, and in any case the day–night variation would make the average over time very small. Some energy is lost upwards by convection and by water condensation, as outlined earlier. The remainder is radiated at planetary wavelengths. The atmosphere absorbs some of this radiation and if it is more transparent at solar wavelengths than at planetary wavelengths then to achieve equilibrium the average surface temperature will rise above what it would have been had the atmosphere been equally transparent over both wavelength ranges.

In the case of the Earth the greenhouse effect results in average surface temperatures that are 35 K higher than would otherwise be the case, the average being 288 K rather than 253 K. The greenhouse effect is produced by the combined action of carbon dioxide, water vapour, and aerosols. You can see from Figure 2.14 that all of these are *minor* atmospheric constituents (aerosols are less abundant than CO_2) and yet without them our planet would not be habitable by its present biosphere, including ourselves.

2.2.4 Climatic change

There are several strands of evidence which strongly indicate that the Earth's climate has changed throughout the whole history of the Earth. The broad picture is one of cooling by about 20 K to the point where, about 800 Ma ago, a series of ice-ages occurred called the late pre-Cambrian series as shown in Figure 2.19 the Cambrian being the name given to an interval of time stretching from about 570 Ma ago to about 500 Ma ago. An *ice-age* is characterized by a cooling of mid latitude and polar regions to the point where polar ice caps, and any small deposits of ice elsewhere, spread to give *glacial conditions*. Within a particular ice-age the glacial conditions can be moderated to give *interglacial conditions* during which the ice deposits retreat to some extent. At present the Earth is in an ice-age (Figure 2.19), though from about 13 000 years ago (= 0.013 Ma) the Earth has been in an interglacial period. By contrast about 20 000 years ago most of the northern

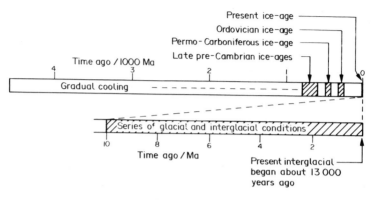

FIGURE 2.19
The Earth's ice-ages (simplified).

hemisphere north of about latitude 50° was covered year around by ice and snow. Tropical latitudes were subject to less cooling. The glaciation of non-tropical regions could begin to return at any time, perhaps taking only a few decades to spread to maximum extent.

Many explanations of the ice-ages have been advanced. Currently fashionable is an explanation first put forward by the Russo-German climatologist Wladimir Peter Köppen (1846–1940) in 1922. It is based on calculations performed in 1920 by the Serbian geophysicist and astronomer Milutin Milankovitch (1879–1958) of the effects of small variations in the Earth's orbital elements and axial inclination on solar radiation falling on the Earth. These orbital and axial variations are largely the result of the gravitational action of the Sun and the Moon on the slightly non-spherical mass distribution of the Earth. Their effect on the solar radiation falling on the Earth is to change its seasonal variability and also to change its variation with latitude. Köppen suggested that these variations in solar radiation were the causes of the ice-ages, though, with some justification, the theory is usually referred to as the *Milankovitch theory*.

It seems likely that these orbital and axial variations are indeed contributory factors, though their importance is not certain. Other likely contributory factors are the general cooling of the Earth over most of Earth history, possible reasons for which will be outlined in section 2.4, the distribution of the continents as determined by Plate Tectonics, slight variations in solar radiation and many others.

In the last 13 000 years there have been several climatic changes a good deal less spectacular than the transitions between glacial and interglacial conditions, but nevertheless of a magnitude that if repeated today would cause considerable disruption to our society. Some of these changes may have been wrought by human activity in the past, particularly forest clearance, which influences the CO_2 content of the atmosphere. It is likely that humanity today is an even greater and growing influence on climate, for example through burning fossil fuels which, as you shall see in the next section, raises the CO_2 content of the atmosphere. Unfortunately the Earth's climatic system is, at present, too poorly understood for the climatic changes of the last 13 000 years to be understood and also for the likely future consequences of our present actions to be foretold.

2.3 The biosphere

The assemblage of all things living, and of their remains, is called the *biosphere*. It has even been suggested that in some senses this is one large organism called *Gaia* of which you and I are individual "cells", though such an idea is highly speculative.

The biosphere contains a great variety of life forms, but they all share many important characteristics two of which are of immediate importance. First, all life forms are able to build up and sustain their bodies, that is, their *tissues*, the basis of which are a variety of *carbon compounds* (section 2.1.1). Second, they all have a source of energy to sustain life, that is, energy for *metabolism*. There are many different ways of tissue building and of metabolism.

A particularly important means of tissue building is *photosynthesis*. A large fraction of present organisms use this method, notably blue-green algae, which are tiny sea-creatures, and green plants on land and sea. Photosynthesis is a process whereby an organism uses solar radiation to turn carbon dioxide and water, plus traces of other compounds, into tissue. A byproduct of this process is oxygen which is released from CO_2 into oceans and atmosphere. Other organisms use sources of energy other than solar radiation to build tissue from simple molecules, whilst others, including ourselves, derive their tissues by eating the tissues of other organisms.

A particularly important means of metabolism is *respiration*, in which an organism takes in oxygen from the atmosphere, or from where it is dissolved in the oceans, and this oxygen then combines chemically with some tissue, releasing energy in the process. Most photosynthesizing organisms also respire, and so do a large variety of other life-forms including ourselves. Respiration reverses the effect of photosynthesis by converting tissue back into water, carbon dioxide and traces of other substances.

When an organism dies its remains may be eaten, or its tissues may combine with oxygen as in respiration and thus more completely reverse the effect of photosynthesis.

Photosynthesis is the major source of oxygen in the oceans and atmosphere today. Therefore, if there was no more to the oxygen story then the present amount of oxygen in the oceans and in the atmosphere could not much exceed that released in the formation of the tissues of organisms alive today and the remains, at the land surface and in the oceans, of dead organisms that have not yet been fully recombined with oxygen. However, there is far too much oxygen in the oceans and atmosphere for this to be the case. This excess oxygen is largely the result of the burial, by geological activity, of a tiny fraction of each of the many generations of remains before they were fully recombined. Geological activity also *exposes* such graveyards, but the balance between exposure and burial was achieved only when such graveyards had become sizeable.

These graveyards include the deposits of coal, oil and natural gas, which are called *fossil fuel deposits*, though most of the buried remains are found in much less concentrated deposits, too dilute to justify commercial exploitation. Thus, the burning of all fossil fuels, which would fully recombine them with oxygen, would not appreciably lower the oxygen content of the atmosphere though the amount of CO_2 released would cause significant climatic change.

It is not only on oxygen that the biosphere has had a great influence. Were it not for biospheric activity then much of the nitrogen in the atmosphere (Figure 2.14) would become incorporated in nitrates and nitrites in the oceans, which at present bear only a small amount of nitrogen (Figure 2.14). Also, small organisms in the oceans greatly accelerate the rate at which CO_2 is removed from the bicarbonate form in the oceans into the less volatile carbonate form ultimately to give substances such as chalk, limestone, marble and coral, which in Figure 2.14 are included in sedimentary deposits. The removal of bicarbonate from the oceans results in the removal of CO_2 from the atmosphere which enters the oceans to replace the lost bicarbonate. This has resulted in a lower CO_2 content of the atmosphere than would have otherwise been the case.

The graveyards of organisms are equivalent to about ten times more oxygen than resides in the atmosphere and oceans today. Most of the remainder, as outlined in section 2.2, has become incorporated in oxygen-poor rocks exposed by geological activity. An appreciable amount of the atmospheric and oceanic oxygen has also gone to form carbonates, which by a roundabout way requires that an oxygen atom be added to CO_2 to form the CO_3 group.

2.3.1 The origin and evolution of the biosphere

Most biologists believe that life originated on the Earth. A small number of scientists believe that life was brought to the Earth in the form of tiny creatures borne, for example, in comets. An intermediate view is that bodies such as comets brought to the Earth highly complex molecules from which it was a relatively short step to a living organism.

Most theories give a central role to oceans, rivers and lakes, and there is abundant evidence that for a long time life was confined to such bodies of water. Such bodies offer the advantage of concentration of the molecular precursors of life, particularly in *shallow* bodies of water to where the precursors could be brought by various means including washing by rain from the atmosphere. Concentration increases the chance of molecular interaction and therefore of the build up of a living organism. Bodies of water also provide a haven for complex molecules which would be readily broken up in the atmosphere or on dry land, and this advantage of a watery habitat also applies to tiny creatures brought to the Earth from space.

There is fossil evidence that life existed in the oceans by no later than about 3500 Ma ago. The first source of materials to build tissue could well have been the remaining molecular precursors. However, the non-biological production rate of such precursors would have been low and a starvation crisis would have loomed. At some time photosynthesis must have been invented and this was a major bulwark against starvation. At first the oxygen released would have been largely confined to the oceans, and much of it would have combined with slightly oxygen-poor rocks on the sea floors and with the sediments washed off the land. But gradually as life spread and the production rate of oxygen grew the oxygen content of the oceans and the atmosphere began to rise.

It is thought that photosynthesis was "invented" about 3200 Ma ago, though this date is very uncertain. At that time the oxygen content of the atmosphere was no more than about 0.1% of its present value and may well have been far less. Most of this oxygen would have been the result of the photodissociation of water by solar uv radiation high in the atmosphere and the subsequent escape to space of hydrogen. There is geological evidence that the amount of oxygen in the oceans and in the atmosphere began to grow appreciably from about 2000 Ma ago and that by no later than about 270 Ma ago had very nearly settled into its present value, though during the preceding 100 Ma or so it may have varied significantly above and below this value in response to variations in the biosphere.

There is abundant evidence that since no later than about 200 Ma ago the amount of oxygen in the atmosphere has hardly varied at all. The amount of a volatile in a repository, as pointed out earlier, is constant if the rate at which it enters the repository equals the rate at which it leaves the repository. These rates often depend on the amount of the volatile in the repository in question and in all the other repositories. Clearly this can lead to very complicated balances. In the case of atmospheric oxygen most scientists believe that the constancy of the amount over the last 200 Ma is largely because the rate of gain of oxygen corresponding to the burial of the remains of dead organisms has almost exactly equalled the rate at which oxygen has combined with newly exposed rocks. A few scientists believe otherwise, namely that certain bacteria have been responsible, bacteria which are known to produce methane (CH_4). This reacts with atmospheric oxygen to yield carbon dioxide and water, and the reaction is so rapid that only a tiny trace of methane exists in the atmosphere. However, the production rate of methane by these bacteria could be sufficiently high that the methane can remove oxygen at a small but significant rate. Therefore, *if* these bacteria respond to an increase in the amount of atmospheric oxygen by increasing their methane production rate, and vice versa, then they could keep the amount of atmospheric oxygen constant. At present there are insufficient data to decide between the majority and minority views.

In the *evolution* of the biosphere there are several general trends. For example, the complexity of organisms has increased from simple single-celled creatures to large multi-cellular plants and animals, though without the disappearance of simpler types. The appearance of oxygen may well have played a crucial role in this, because respiration may be the only suitable source of metabolic energy for multi-cellular organisms. Also, there has, on the whole, been an increase in the number of different *species*, that is, types of organism. However, there have also been a few relatively sudden and substantial *decreases* in the number of species. For example, about 65 Ma ago over half of the previously existing species of multi-cellular organisms disappeared, among them the dinosaurs. Many explanations have been advanced for such catastrophes, including climatic changes, which in some explanations are wrought by the collision of large extraterrestrial bodies with the Earth. However, the truth of the matter remains a mystery.

There is abundant evidence that all life forms on the Earth today have a common origin. This is apparent at molecular level where, for example, we are virtually indistinguishable from dogs, trees and bacteria. It is also apparent on a larger scale, from cells to eyes to skeletons. Moreover, a fairly complete *evolutionary* sequence can be established from the origin of life to today. The *process* of evolution is not well understood, but the idea that evolution has actually occurred is very well founded on abundant evidence.

The time-span of the development of the biosphere is the age of the Earth, about 4600 Ma. Imagine that this is represented by one year of our time, and that the Earth is created on 1 January. The first life-forms, single-celled organisms, appear no later than the end of March. Photosynthesis is "invented" in late April, but multicellular organisms do not appear until late September. There is no life on the land until the end of November. The first primitive human beings do not appear until about 10:00 hours on 30 December and our own species not until about 23:03 hours on 31 December. Written history begins at about 34 seconds to midnight on 31 December, and Christ lives at about 14 seconds to midnight. I am impressed by how recently our species appeared, and by how long the land was utterly devoid of life.

2.4 The evolution of the Earth's atmosphere and surface volatiles

It is clear from the preceding sections that the nature of the Earth's present atmosphere is largely the result of biospheric activity, and that the biosphere has also had a significant effect on the surface volatiles.

It is possible to infer what the atmosphere would have been like before the biosphere had had much effect. Much of the nitrogen, the dominant constituent of the atmosphere today, would have resided in nitrates and nitrites in the oceans. Almost all of the oxygen would have resided in CO_2, the trace of atmospheric oxygen being largely the result of the photodissociation of water. There would have been some carbonate formation, mainly in the oceans, where even in the absence of life the rate of formation would have been higher than on land. However, the overall rate would have been far lower and therefore the amount of CO_2 in the atmosphere and the amount of bicarbonate in the oceans would have been higher than today. The extra oxygen atom required to turn CO_2 into CO_3 would ultimately have come, as today, from the atmosphere.

At that distant time water, as today, would be the most abundant volatile and would largely reside in the oceans, the biosphere having had little effect on the copious quantities of water. If there is a surface deposit of water, such as oceans, then the greater the surface temperature the greater the amount of gaseous water in the atmosphere. Thus there may then have been either more or less water in the atmosphere than there is today.

Argon is also largely free from biospheric influence and would have been present in comparable abundances to today unless it was then locked up in igneous rocks to a different extent.

Thus, before the biosphere had had much effect, the atmosphere would have consisted largely of carbon dioxide, water and nitrogen. The total mass of the atmosphere was probably less than it is today. The dominant surface volatile would have been water, mainly in the oceans as today, and in these oceans there would have been considerable quantities of bicarbonate, nitrates, nitrites and other volatiles.

Evidence for this conclusion is provided by the gases which all volcanoes emit (in addition to lava and ash). Water is the most abundant gas, followed by CO_2, oxides of sulphur, nitrogen and smaller quantities of other substances including argon. The amount of oxygen is negligible. Though most of these gases are recycled from the various surface repositories of volatiles, they have been freed of biospheric influence. Note that the oxides of sulphur, to which I have paid little attention, largely dissolve in the oceans, and subsequently form sulphates which largely reside in sedimentary rocks (Figure 2.14). The biosphere has had a relatively small effect on the sulphur story.

There are, however, two main problems that have to be faced by the prebiospheric atmosphere at which we have arrived.

First, you will recall from section 2.2.4 that the Earth's surface temperature was probably higher in the distant past than today. Also, there is ample geological evidence that *liquid* water has been present at the Earth's surface ever since well before 3000 Ma ago. Therefore the temperature over much of the Earth's surface has been above the triple-point value of 273 K throughout this time. However, theories of stellar evolution, which are well founded on observations, indicate that 4600 Ma ago the Sun's luminosity was only about 70% of its value today and that the luminosity has since increased to its present value over a protracted period. If the Earth in its present state were transported back about 4600 Ma then the surface temperature would everywhere fall well below 273 K, the oceans would freeze, and any land would be covered with snow and ice. In this ice-bound state the Bond albedo of the Earth would be so high that the Earth would probably have remained ice-bound until about 2000 Ma ago, and perhaps even more recently, in contradiction to the geological evidence.

In order to get around this difficulty it is necessary for the early atmosphere to have contained at least 10 to 100 times more CO_2 than it does today. The greenhouse effect would then have been sufficiently large to have prevented the freezing of water. This amount of CO_2 is far less than the present amount of oxygen in the atmosphere, and as this oxygen represents only a fraction of the biospherically disrupted CO_2 it may seem that there is no difficulty. However, it's not quite as simple as this because CO_2 readily dissolves in the oceans, to a far greater extent than oxygen. Nevertheless, it is certainly plausible that CO_2 could have been sufficiently abundant in the atmosphere to prevent freezing, particularly if the Earth acquired its atmosphere and surface volatiles fairly rapidly.

Second, if life originated on the Earth, then it is easier to see how it could have happened if the atmosphere contained hydrogen-rich molecules such as methane (CH_4) and ammonia (NH_3) rather than oxygen-rich molecules such as CO_2. (Note that in this respect water is neutral, being both hydrogen rich and oxygen rich.) This conclusion rests on laboratory work in which gas mixtures have been exposed to the kinds of energy sources that would have been available in the early Earth atmosphere, such as solar uv radiation (there was little O_2 and therefore negligible amounts of uv absorbing ozone), and electrical discharges (such as lightning). When the gas mixtures contain CH_4 and NH_3 then comparatively large quantities of the molecular precursors of life are produced. Far smaller quantities of such precursors are produced when the gas mixtures contain CO_2 and N_2 rather than CH_4 and NH_3.

This has led some astronomers to propose that the early atmosphere did indeed contain CH_4 and NH_3 rather than CO_2 and N_2. Moreover, a mere trace of NH_3 would have prevented the Earth's water from freezing, because

NH_3 is very efficient at producing a greenhouse effect.

However, the origin of the precursors on Earth cannot be ruled out in an atmosphere in which the abundances of CH_4 and NH_3 are negligible, particularly if traces of CO and H_2 are present, and CO and H_2 are among the trace gases emitted by volcanoes today. Moreover, it is just possible that the precursors originated elsewhere than on Earth. In both cases the reasons for preferring an early atmosphere containing significant amounts of CH_4 and NH_3 are weakened.

But even if the early atmosphere did contain CH_4 and NH_3 these gases would inevitably revert in the main to CO_2 and H_2 respectively. This reversion would result from the photodissociation of CH_4 and NH_3 by solar uv radiation, the hydrogen liberated escaping from the exosphere. The reversion would be aided by small quantities of oxygen, for example from the photodissociation of water, which would chemically react with CH_4 and NH_3, removing hydrogen to form water molecules. Some of this oxygen would also combine with the liberated carbon (C) to form CO_2, though much of such oxygen would come from crustal rocks, chemically acted on by carbon. The early biosphere may also have aided the removal of any CH_4 and NH_3 by using it as "food" for tissue building.

Any such reversion must have been complete by about 2000 Ma ago at the very latest, because by then the abundance of atmospheric oxygen was appreciable and growing and this is inconsistent with the presence of appreciable quantities of CH_4 and NH_3 which would readily react with oxygen and thus remove it from the atmosphere.

The *origin* of the Earth's atmosphere and surface volatiles is a subject for Chapter 5, and their distant fate and that of the biosphere is a subject for section 15.6.

2.5 Summary

The model of the Earth's interior, summarized in Figure 2.3, is consistent with the constraints supplied by external observations and by theory. The broad features of the model, the division of the Earth into a core consisting largely of iron and subdivided into an inner (solid) and outer (liquid) core, and a largely solid mantle and crust consisting largely of oxides and silicates, are almost certainly correct. But the detailed features (not shown) may not be, such as the minor constituents of the core and the details of mantle composition.

The Earth's lithosphere is very geologically active. Many of its features and the processes that mould it are explained by the theory of Plate Tectonics in which the lithosphere is divided into several fairly rigid plates which are in motion with respect to each other and which are being destroyed at some boundaries and created at others. It is likely that in spite of its largely solid-state thermal convection in the mantle plays an important role in plate motion.

The high level of geological activity of the Earth is probably sustained today largely by radioactive heating throughout much of the interior, though in the distant past the heat of accretion and perhaps heat from short-lived radioactive isotopes may have been important.

The Earth was probably formed from dispersed matter about 4600 Ma ago. After at most a few hundred Ma it had acquired an atmosphere (the origin of which is a subject for Chapter 5) which probably consisted largely of carbon dioxide, nitrogen, and water, and it had also acquired extensive oceans of water. Not long afterwards the biosphere was established and very gradually it transformed the Earth's atmosphere to the oxygen-nitrogen atmosphere we possess today. Subsequently the biosphere grew in complexity and comparatively recently gave rise to a life-form, ourselves, which is able to contemplate from where it came and to where it is going and which is now able to significantly influence climate and the atmosphere.

2.6 Questions

1. In a couple of sentences, say why it cannot be *proved* that the Earth is like the model shown in Figure 2.3.

2. It is proposed, by Dr. Lackof D. Ata, that the Earth consists of a core of radius 3400 km and mean density 11 000 kg/m^3 the remainder of the Earth being of mean density 5000 kg/m^3. Show that this model *cannot* represent the Earth.

3. List the different types of evidence which help us to elucidate the nature of the Earth's interior (including the lithosphere) and indicate which features of the interior are elucidated in each case.

4. Contrast the source of the gravitational field of the Earth with the likely source of its magnetic dipole field, and contrast also the configurations of the field lines.

5. For the model of the Earth in this Chapter give examples of divisions of the Earth's interior based on (a) composition, (b) phase, (c) plasticity. Outline the role played by *temperature* in cases (b) and (c).

6. State the differences (if any) between an element, an isotope, an ion, a molecule, a chemical compound, a pure substance, a mineral, a mixture, a substance, a material and a rock.

7. In a certain chemical reaction two molecules A and B join to make a third molecule C. Under what conditions of pressure and temperature would you expect the production of C to be favoured?

8. Here are the stages in the formation of a certain rock. (a) Magma solidifies beneath the Earth's surface. (b) The solidified magma is subsequently exposed, and through the action of wind, rain and frost is

broken up and modified chemically. (c) A stream transports the smaller fragments and deposits them in a lake. (d) The fragments near the base of the deposit are ultimately compressed by the overburden to form a material of rock-like hardness. (e) A nearby source of magma heats this material near to melting, though it does not melt. It subsequently cools, and now we possess a piece of it.

Name the sequence of rock types through which the material passes and the processes that transform one into the next.

9. A rock, known to have solidified from a molten state, and not appreciably heated since, contains a ratio of ^{40}Ar to ^{40}K of 1:9. How long ago did it probably solidify? What assumptions do you make? In terms of atomic constitution, what is the significance of the symbols Ar, K and the number 40?

10. List six large-scale topographic features of the Earth's surface which are explained by the theory of Plate Tectonics.

11. In what way does *heat* differ from other sources of energy?

12. List possible sources of heat for the Earth's interior, and in each case state if the importance of the source depends on whether the Earth was accreted heterogeneously or homogeneously.

13. Contrast thermal *conduction* and thermal *convection*, and describe their possible roles in relation to the motion of lithospheric plates and to the thermal evolution of the Earth.

14. A substance has a triple-point temperature of 200 K. Why is this substance regarded as a volatile? What further information is required in order to determine whether it would be a liquid at the present mean surface temperature of the Earth?

15. Justify the adjective "surface" in surface volatiles.

16. List various repositories of oxygen in *volatile* form on the Earth, and outline some ways in which these repositories lose and gain oxygen.

17. Sketch the spectrum of a thermal source of em radiation at a temperature of 5000 K, and indicate the wavelength ranges corresponding to uv, optical, visible, solar, infrared, planetary and radio waves.

18. Sketch the pressure and temperature versus altitude in the Earth's atmosphere. Indicate the altitudes at which clouds, plasma and variations in composition occur.

19. Outline some of the major factors which determine the circulation of the Earth's atmosphere.

20. Explain why
    (a) seasonal variations at the equator are likely to be less than those at higher altitudes,
    (b) mid-winter in the southern hemisphere is in mid-June.

21. A certain gas transmits 50% of em radiation at solar wavelengths and 75% of em radiation at planetary wavelengths. Why would such a gas *not*

give rise to a greenhouse effect?

22. What is the Bond albedo of a planet which absorbs 15% of the solar radiation incident upon it? Could *local* albedos be different?

23. What is climate, and outline some ways in which it can be caused to change?

24. Summarize the effect of the biosphere on the atmosphere of the Earth, and the evidence for and against an early atmosphere of CH_4 and NH_3.

3

MARS

The night sky is unfamiliar to many people today because it can hardly be seen through the brightly lit haze that hangs at night above most cities and large towns, where many people live. Thus, when Mars goes through an opposition it is largely unnoticed. But until a few generations ago it was plain for all to see, tinted red and very bright. "Mars" is the name of the Roman god of war, an appropriate association for a red-tinted planet.

The motions of the planets but of Mars in particular played a crucial role in the development of astronomy and thus of science as a whole. Between 1576 and 1597 the Danish nobleman and astronomer Tycho Brahe (1546–1601) determined the apparent path of Mars against the stars far more accurately than had previously been achieved. In 1600 Tycho acquired a new assistant, the young Johannes Kepler. Kepler spent a long time trying to understand the apparent path of Mars, and in 1609, some years after Tycho's death, he published his first two Laws of Planetary Motion (see section 1.1).

The path of Mars led to these laws for two reasons. First, it orbits the Sun *beyond* the Earth and is consequently readily observed over much of its orbit, unlike Mercury and Venus which orbit the Sun *within* the Earth's orbit and therefore are drowned in the light of the Sun for much of the time. Second, of the then known planets beyond the Earth (Uranus, Neptune and Pluto were still undiscovered) Mars has the most eccentric orbit. Therefore, it was easier to discover the elliptical nature of planetary orbits (Kepler's first Law) and the way a planet moves around the ellipse (Kepler's second Law) from the path of Mars than from the path of any other planet.

Kepler added his third Law nine years later, and in the latter half of the seventeenth century Isaac Newton explained Kepler's three Laws by theories of gravity and motion as outlined in Chapter 1.

The exploration of Mars itself, as opposed to its orbit, began in 1609 when the Italian scientist Galileo Galilei (1564–1642) turned the newly invented telescope to the skies.

3.1 The exploration of Mars

Galileo saw Mars as a disc rather than as a point of light and at that time this was an important discovery in itself because it suggested that Mars was a world. But it was not until 1659 that the first clear drawings of surface markings were made. These are by the Dutch scientist Christiaan Huyghens (1629–1695) who drew a large roughly triangular feature, dark on the otherwise red surface. There is little doubt that this is the feature now called Syrtis Major (section 3.3). Moreover, Huyghens was able to follow the dark feature night by night and recorded in his dairy on 1 December 1659 that

"The rotation of Mars seems to take 24 terrestrial hours like that of the Earth."

He was very nearly right as Table 1.3 shows. The axial rotation period of Mars is called the *sol*. The *mean solar sol* is now known to be 24h 39m 34.7s, and the *sidereal sol* is 24h 37m 22.7s. The sidereal *orbital* period of Mars is a good deal longer than that of the Earth, 686.98 mean solar days rather than 365.26 mean solar days, and therefore there are 668.60 mean solar sols in the sidereal orbital period of Mars.

The year 1877 was a remarkable one in the history of the exploration of Mars. In that year there was a particularly favourable opposition, that is, the distance between the Earth and Mars was about as small as it can ever be. The opposition distance varies from 55×10^6 km to 102×10^6 km, mainly because of the comparatively large eccentricity of the orbit of Mars (see Figure 1.6) the closest oppositions being when Mars is near perihelion. The apparent diameter of Mars correspondingly varies by about a factor of 2. The apparent diameter is measured by the *angle* the object subtends at the observer as shown in Figure 3.1. This is called the *angular diameter*. The

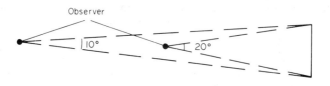

FIGURE 3.1
Angular diameter.

angular diameter of the Moon is about half a degree (°), and as there are 60 *arc-minutes* (′) in a degree this is about 30′. The angular diameter of Mars at opposition can vary from 0.42′ to 0.23′. In 1877 the angular diameter of Mars at opposition was 0.410′, and as there are 60 *arc-seconds* (″) in an arc-minute this is 24.6″.

By 1877 many dark markings had been mapped on the surface of Mars and changes in these markings had been extensively studied. There were changes in shape, extent and in contrast against the comparatively light red tint of the rest of the Martian surface. Some of these changes followed the seasons. Most astronomers believed that the dark areas were seas and the light areas continents (see Figure 2.1) and they had been named accordingly. A few astronomers believed that the dark areas were vegetation, perhaps filling basins that had once been seas.

It had long been known that Mars possessed white polar caps and that these advanced and retreated with the seasons in the appropriate manner. It was known, from observations of the apparent movements of surface markings as Mars spins on its axis, that the axial inclination was about 24°, and therefore the seasonal variations in solar radiation are comparable with those on the Earth, though the greater eccentricity of the Martian orbit significantly enhances the seasonal variation in the southern hemisphere where mid-summer occurs soon after perihelion. This is why the southern polar cap advances and retreats with the seasons to a far greater extent in the southern hemisphere than in the northern hemisphere. The polar caps were widely regarded as consisting of water, condensed as ice and snow as on the Earth.

There was ample evidence by 1877 for a Martian atmosphere. The polar caps could not exist without an atmosphere: they would rapidly evaporate and the water would escape to space, and this would also apply to any open bodies of liquid water. Moreover, as early as 1809 yellow clouds had been observed, and by 1858 small white clouds had been seen also.

There were no measurements of temperatures at the Martian surface, but calculations, laced with a heavy dose of speculation, gave values similar to the Earth's surface temperatures.

On such an apparently warm moist planet the possibility of life seemed well worth entertaining. Indeed, several schemes had already been proposed for signalling to the inhabitants of Mars from the Earth.

This was the position as the opposition of 5 September 1877 gradually approached. Advances in optics and in scientific equipment in general meant that the astronomical community was considerably better prepared than it had been for earlier favourable oppositions. It is not hard to imagine the excitement mounting as Mars brightened and its angular diameter grew.

Throughout the weeks surrounding the opposition the Italian astronomer Giovanni Virginio Schiaparelli (1835–1910) scrutinized Mars at the Brera Observatory Milan where he was director. He made extensive observations with a telescope with a main lens 220 mm diameter, this being an important determinant of the smallest detail that could be seen (see section 3.1.2). By the standards of the day this was a large telescope.

His maps of Mars were the best yet and we still use the names he gave to the various dark features. But he is best remembered for about forty fine lines that he drew crossing the bright red areas, *canali* as he called them. The Italian word *canali* means grooves but drop the "i" at the end and you have a sensation. Description becomes interpretation and in a climate of opinion which considered life on Mars a reasonable possibility it was not entirely

ridiculous to imagine that any such life was intelligent, and that it had built
canals.

And yet, those scientists who at first took to the notion of "canals" were by
and large *not* astronomers. Until the 1880s Schiaparelli alone had seen them,
and most astronomers simply did not believe in the existence of fine lines on
Mars, regardless of their interpretation. This is not quite as silly as it sounds.
Mars is a small planet, about half the diameter of the Earth, and even during
favourable opposition is only about 0.4′ in angular diameter. It requires
visual acuity, skill, experience and a good telescope to see anything much on
the surface at all. It also requires good so-called "*seeing*", that is, a clear
steady Earth atmosphere above the telescope. Here we have three ingredients: experimental skill; good apparatus; luck. These ingredients have been
and continue to be vital to progress in experimental science. For example, at
first only Galileo could see four satellites orbiting Jupiter. They later proved
to be real enough so why not the canali? Indeed, by the next favourable
opposition, that of 1892, Schiaparelli was not alone in having seen the canali.
Other astronomers had seen them and a few of the more prominent canali had
even been identified on maps earlier than 1877.

Nevertheless, the majority of astronomers believed that Mars bore *no*
canali. The minority which believed that canali existed divided into those that
thought the canali were natural, and those that thought they were artificial.

At first, many of those that thought they were natural considered the canali
to be water channels, joining one sea to another across the red continents. But
the idea that the dark areas were seas crumbled within a few years of 1877.
The "seas" did not reflect light in the right sort of way. Then there were the
changes in extent, shape and contrast, for which no convincing explanations
had been devised if the dark areas were seas. Moreover, the dark areas were
not featureless — structures within them had been seen. They were even
crossed by some of the canali. The interpretation of the dark areas therefore
shifted to vegetation and the canali were thought to be some other natural
feature such as a tract of vegetation. Schiaparelli started off by believing the
dark areas to be seas, the canali to be water channels, but towards the end of
his life (he died on 4 July 1910) he too had changed his mind and considered
the dark areas to be vegetation the canali to be natural vegetation tracts.

In the same way, those that thought the canali were artificial considered
them at first to be water channels, but as the interpretation of the dark areas
shifted from seas to vegetation then the canali were considered to be irrigated
tracts of land, that what was seen was vegetation sustained by a thread of
water too narrow to be seen. Indeed, those that regarded the canali as
artificial greatly preferred the vegetation hypothesis for the dark areas and
were early supporters of this view. This is because a canal network could then
be understood as an attempt to distribute a meagre Martian water supply,
hardly meagre if the dark areas were seas.

The idea that Mars was short of water stemmed from the interpretation of the red areas as deserts and from the rapidity with which the polar caps advanced and retreated with the seasons, thus indicating that the caps were thin. (The absence of seas would not necessarily indicate a dry planet.) Moreover, a dry Mars fitted a now obsolete theory of planetary evolution one feature of which is that the planets formed in order, starting with Pluto and ending with Mercury. Thus, Mars would be older than the Earth and hence would have been cooling down longer and losing its water for longer by escape into space. And not only would Mars have had longer than the Earth to lose its water, its lower surface gravity would also have made escape easier.

A dying world then, cooling down, and becoming desiccated. What more natural than for its inhabitants to build canals?

In the 1890s there emerged two powerful supporters of the view that the canali were associated with canals. One was the American astronomer William Henry Pickering (1858–1938) who began to observe Mars in 1892. The other was another American astronomer, and perhaps the man most associated with the canals, Percival Lowell (1855–1916), who in 1894 founded an observatory at Flagstaff in Arizona, mainly for him to study Mars.

Soon afterwards the world of literature stirred in response to the canali. In 1897 appeared *The War of the Worlds* by the British writer Herbert George Wells (1866–1946), one of the finest narratives ever written and in which the Martians look Earthwards and see

"... a morning star of hope, our own warmer planet, green with vegetation and grey with water, with a cloudy atmosphere eloquent of fertility, with glimpses through its drifting cloud wisps of populous country, and narrow, navy crowded seas"

and we are destined to be invaded.

A fine tale, and scientifically speaking Wells' story in the late nineteenth century was credible enough. And yet some of its immediate popularity arose not from the public's fear of Mars but from the then current fears in Britain of invasion by Germany. It seems likely that some of those early readers saw, not Wells's Martians rampaging through England, but the Kaiser's army.

Wells's story is among the first in a long line of science fiction stories involving Mars. Little of it rises to the standard set by Wells.

But good or bad this science fiction of Mars kept alive in the public mind for some decades into this century the idea that intelligent life existed on Mars. By contrast, in the scientific community as the years went by evidence mounted against the idea that any such life existed. Measurements showed that Mars was not as hospitable as had been thought at the end of the nineteenth century, but is more harsh than a dry antarctic desert. And the long clement nurturing needed for the development of advanced forms of life, leave alone intelligent life, did not seem likely to have happened on Mars.

Nevertheless the canal controversy persisted into the era of the Space Age in spite of the fact that ever since their discovery in 1877 there had always been a majority of astronomers who claimed that the canali simply did not exist, that there was no basis for even asking the question whether they were natural or artificial. Only the Space Age itself brought the final solution. Since 1965 many spacecraft have been to Mars. Figure 3.2 is a spacecraft picture of Mars showing rather more detail than ever seen from the Earth. No canali can be seen. Nor can any be seen in far more detailed pictures. *The canali do not exist.*

FIGURE 3.2
Mars imaged by Viking 1 at a range of 560 000 km. Olympus Mons is the dark spot at top right. Much of the brightness variation across the disc arises from changing illumination angle rather than from albedo variations, but there are some bright whitefall patches. The north pole is over the horizon at top right.

What then are we to make of Lowell in "Mars as the abode of life" (1908),

> ". . . not only do the observations . . . (on the canals) . . . we have scanned lead us to the conclusion that Mars at this moment is inhabited, but they land us at the further one that these denizens are of an order whose acquaintance was worth the making."

Or even of Schiaparelli (in the journal *Natura ed Arte* (1892)):

> "It is not necessary to suppose them to be the work of intelligent beings, and notwithstanding the almost geometrical appearance of all of their system, we are now inclined to believe them to be produced by the evolution of the planet, just as on the Earth we have the English Channel and the Channel of Mozambique."

The explanation seems to be that the canali *were* evidence of intelligent life, but, as the American astronomer Carl Sagan (b. 1934) has put it, the intelligence was at the *eyepiece* end of the telescope. The human mind, straining to interpret elusive detail at the limit of perception, drew order from chaos and invented narrow linear features which are simply not there. The canals stand not as a chronicle of Mars but as a monument to the subtleties of human visual perception.

3.1.1 Phobos and Deimos

The opposition of 1877 also yielded two satellites of Mars.

In the months leading up to the opposition the American astronomer Asaph Hall (1829–1907) was searching for Martian satellites using a telescope with a main lens 660 mm diameter. The failure of earlier searches by others indicated that any satellites would be faint. Indeed, Hall's search very nearly failed because by early August he seems to have been near to giving up. However, some wifely encouragement made him continue and early in the morning of 11 August he suspected he had discovered a satellite. Overcast skies delayed confirmation until 16 August. And then on 17 August he discovered a *second* satellite. Hall soon named the satellites.

> "Of the various names that have been proposed . . . I have chosen those suggested by Mr. Madan of Eton, England, viz. Deimos for the outer satellite: Phobos for the inner satellite."

In mythology Deimos and Phobos are the horses that drew the chariot of Mars. "Deimos" means "flight" and "Phobos" means "fear". Deimos was the first to be discovered.

These are *tiny* satellites. Phobos is about 11 km mean radius and Deimos about 6 km mean radius. Their orbits are of fairly small eccentricity and of low inclination with respect to the equatorial plane of Mars. The semi-major axes are also small, that of Deimos is 23 500 km and that of Phobos is 9380 km. Thus, Phobos lies only about 6000 km above the surface of Mars and cannot be seen from high latitudes on the Martian surface.

The semi-major axes and orbital periods of Phobos and Deimos enabled the mass of Mars to be determined for the first time, by a technique outlined in section 3.2.1.

Until the Space Age very little was known about Phobos and Deimos. Figure 3.3 shows photographs obtained in the late 1970s by spacecraft orbiting Mars. You can see that these satellites are irregular battered bodies. They are probably captured asteroids — more on this in Chapter 14.

(a)

(b)

FIGURE 3.3
(a) Phobos imaged by Viking Orbiter 1. (b) Deimos imaged by Viking Orbiter 2.
These are shown to their correct relative sizes. The mean radius of Phobos is about
11 km, and is about 6 km for Deimos.

Appropriately enough, the largest crater on Phobos has been named Stickney, the maiden name of Asaph Hall's wife.

There are two apparently curious anticipations of the discovery of Phobos and Deimos, in the satirical novel *Gulliver's Travels* by the Anglo-Irish writer Jonathan Swift (1667–1745), and in the philosophical fantasy *Micromegas* by the French writer Francois Marie Arouet de Voltaire (1694–1778). However, it seems that these were no more than educated guesses, based on a now outmoded theory of the number of satellites that each planet should have.

● *3.1.2 Information from electromagnetic waves*

To be able to follow the story of Martian exploration into the twentieth century, and for many other reasons, it is necessary for you to know how information about a distant body (such as Mars) is bourne by, and extracted from, electromagnetic (em) waves.

The information is normally collected by telescopes. There are two sorts: *refractors* are based on a large lens as shown in Figure 3.4 (a); *reflectors* are based on a large concave mirror as shown in Figure 3.4 (b). The *area* of the

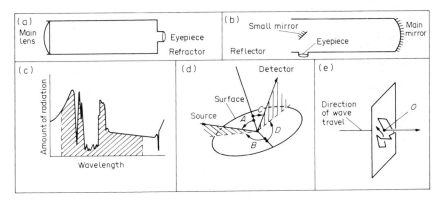

FIGURE 3.4
(a) A refractor. (b) A Newtonian reflector. (c) Radiometry and spectrometry. (d) Photometry. (e) Polarimetry.

lens or the mirror determines the *amount* of electromagnetic radiation collected, and the *diameter* helps determine the *resolution* in the image, that is, the amount of detail. The improved resolution can be made accessible to the eye by means of an eyepiece the overall effect being an apparent increase in the angular diameter of the object. The factor by which it is increased is called the *angular magnification*. The larger the lens or the mirror the better, in most cases. *Radio telescopes*, which work at radio wavelengths (Figure 2.16), are always of the reflecting kind.

But in addition to images there are *four* other types of information borne by em waves.

First, there will be a certain total amount of radiation over some range of wavelengths, as illustrated by the shaded area in Figure 3.4 (c). Measurement of this is called *radiometry*. Bodies in the Solar System (other than the Sun) act as sources of em waves at *solar* wavelengths largely because they *scatter* solar radiation, and we *see* such bodies because of the radiation scattered at visible wavelengths. By contrast, these bodies *emit* nearly all of the em radiation we receive from them at *planetary* wavelengths. Radiometry at planetary wavelengths can reveal surface temperatures, the higher the temperature the greater the amount of radiation from a given surface area as indicated by Figure 2.16.

Second, there will be a detailed distribution of em radiation with wavelength, as in a spectrum (section 2.2.1). *Spectrometry* is the detailed measurement of spectra. Figure 2.16 shows spectra that would correspond to em radiation from the surface of a relatively simple body, such as a lump of iron. You can see that the peak intensity shifts to shorter wavelengths as temperature rises and therefore the position of this peak as determined by spectrometry will yield the surface temperature. Though a planet, or the Sun, has a spectrum rather like those in Figure 2.16 there is also a considerable amount of fine detail. For a small wavelength range this is illustrated by Figure 3.4 (c). The "spikes" and "troughs" in this spectrum can carry considerable information about the body emitting or scattering the em radiation, because a particular atom, molecule, or mineral will "write" a characteristic "signature" on the spectrum. The signature can consist of absorption of certain wavelengths, thus giving narrow troughs in a spectrum, or of emission of certain wavelengths, thus giving narrow spikes in a spectrum, depending on the physical situation. A particular type of atom or molecule can be identified if it is in gaseous form and sometimes if it is in a liquid or solid substance. Sometimes it is possible to tell whether the identified substance is a solid, liquid or a gas.

In the case of a planet, spectral signatures can be impressed on scattered solar radiation or on planetary radiation. The signatures can originate from substances in the atmosphere or at a planet's surface though signatures from a planet's surface are fewer and harder to interpret than those from the atmosphere and the same applies to condensed materials in clouds.

The signatures are sensitive to the physical environment of a substance, and can be used, for example, to obtain the pressure and the temperature.

Third, em radiation is usually emitted or scattered by a body unequally in different directions. The case of scattering is illustrated in Figure 3.4 (d). The amount of radiation scattered by a given area of the body will depend on the angles A and B, and the amount detected will depend as well on the angles C and D. If the radiation is *emitted* by the surface, rather than scattered by it, then the amount detected will depend on the angles C and D alone.

Certain aspects of the nature of a surface can be gleaned from these variations with angle. The measurement and study of such variations is called *photometry*. As a planet spins on its axis a particular area on its surface as viewed from the Earth will be seen at different values of C and D, and the area will be illuminated by the Sun at different values of A and B. As the planet moves around its orbit this will also cause changes in the angles A, B, C, D.

Fourth, the electric and magnetic fields in an electromagnetic wave have directions at right angles to the direction of wave travel rather in the manner of an S wave (Figure 2.6 (b)). Imagine a detector that would visibly respond to an electromagnetic wave, and that such a wave travels in the arrowed direction in Figure 3.4 (e). Before the wave arrives the detector sits at O, and as the wave passes O the deflection of O is in the plane perpendicular to the wave direction. Information about the source of the wave can be extracted from the pattern of the "dance" of the detector. The (indirect) measurement and study of these "dances" is called *polarimetry*.

3.1.3 The exploration of Mars in the twentieth century

The first space-vehicle to visit Mars was the US spacecraft Mariner 4 which flew by Mars in July 1965. Let's summarize the ideas about Mars that were current at that time.

The best sort of Earth-based map of Mars that was then available is typified by Figure 3.10 in section 3.3.2 *provided* that you view it from a distance of about 5 metres. The best maps were made from a combination of photography and visual observations. Photography has the advantages that it yields accurate shapes and that by using exposures of several seconds very low contrast features can be detected. Visual observations have the advantage that glimpses of Mars during brief moments of good seeing yield detail beyond the reach of photography, because during typical photographic exposure times atmospheric turbulence blurs fine detail.

The features visible in Figure 3.10 from about 5 metres are variations in albedo across the Martian surface. If this were in colour then large-scale colour variations would also be apparent. These variations do not necessarily correspond to topographic features. In 1965 the only topographic data were from radar studies (see section 4.1.1). These data were of low resolution, that is, they only revealed topography on a comparatively large horizontal scale. However, they did indicate altitude differences of up to 16 km between different areas of Mars.

The albedo and colour features are subject to the seasonal and non-seasonal changes mentioned earlier. Moreover, there was now evidence that in the hemisphere in which it was spring, and in which the polar cap was consequently warming, the dark areas became even darker. There was even some evidence for a "wave of darkening" spreading from the waning cap towards the equator. These seasonal changes in the dark areas kept alive the idea that the dark areas were vegetation, being revived in the spring by moisture released from the waning cap.

But there was a minority of astronomers which believed that the dark areas were distinct from the bright areas only in being of a different non-biological composition, just as there are rocks on Earth with different colours and albedos. For example, it had been suggested by the American astronomer Dean Benjamin McLaughlin (1901–1965) that the dark areas were volcanic ash ejected from still-active volcanoes and placed in semi-permanent patterns by the prevailing winds in the Martian atmosphere. It had also been suggested that the dark areas contained substances the darkness of which changed as they absorbed water, thus accounting for the seasonal darkening in spring.

The bright areas were regarded as dusty deserts by almost everyone, and there was a comparable degree of agreement that the polar caps consisted of condensed water, in the form of snow or frost, and that the rapid and extensive seasonal advances and retreats of these caps indicated that they were no more than 100 to 200 mm deep. Those who believed that the dark areas were vegetation saw in the variations in their size and shape a battle between the vegetation and the encroaching desert.

Temperatures at the Martian surface had been established by radiometry and by spectrometry and indicated that near midday in the Martian tropics the surface temperatures could rise as "high" as about 280 K, but that near sunrise after a long cold night the surface temperatures were typically 228 K. These large diurnal swings in temperature indicated that the Martian atmosphere was a good deal less effective than that of the Earth in blocking planetary radiation to space at night, and thus allowed the Martian surface to cool considerably. It followed that the column mass of the atmosphere was a good deal less than that of the Earth. Detailed analyses of the solar radiation scattered by Mars, plus some rather uncertain assumptions about the way the Martian surface and atmosphere interacted with solar radiation, led to several estimates that gave an atmospheric surface pressure between about 80 and 120 millibars (1000 millibars = 1 bar), considerably less than the 1000 millibars or so at the Earth's surface. In terms of column mass the difference is less because surface gravity on Mars is only 38% of that on the Earth. On Mars the pressure range corresponds to 0.21×10^4 to 0.32×10^4 kg/m^2 compared to 1.04×10^4 kg/m^2 on the Earth.

The composition of the atmosphere had been investigated by means of spectrometry. Solar radiation is partially scattered and partially absorbed by the Martian atmosphere. The remainder suffers a similar fate at the surface and of that scattered by the surface some is absorbed and scattered again by the atmosphere and some of that which is scattered escapes to space. Therefore, solar radiation scattered by Mars and reaching the Earth will have both atmospheric and surface signatures impressed on it. In practise the atmospheric signatures can be separated thus enabling atmospheric gases to be identified.

Thus, by 1965 it had been known for some years that carbon dioxide was present in the Martian atmosphere and the strength of the signature indicated that it accounted for only a few millibars of the 100 or so total pressure. This is the *partial pressure* of the CO_2. Water vapour had also been identified at the very limit of what could be detected by the techniques then available and it was clear that the Martian atmosphere contained *far* less water than the Earth's atmosphere.

The bulk of the presumed total pressure thus remained unaccounted for. A widely held view was that, as on the Earth, nitrogen was the predominant component. There would be little hope of detecting nitrogen from the Earth, for two reasons. First, nitrogen has a far weaker spectral signature than carbon dioxide. Second, there are copious amounts in the Earth's atmosphere, which readily cloaks the signature from Mars. By contrast, there is comparatively little CO_2 in the Earth's atmosphere. Water, like CO_2, has a strong spectral signature but it shares the second problem with nitrogen that there are comparatively large amounts in the Earth's atmosphere. This problem was overcome by making observations at a high-altitude desert site, above which there was little water in the Earth's atmosphere, and by using the slight shifts in the wavelengths of any Martian spectral features with respect to the wavelengths of such features in Earth-bound substances. These shifts are called *Doppler-shifts* after the Austrian physicist Johann Christian Doppler (1803–1853) and in this case arise from the relative motion of the Earth and Mars.

Oxygen would be as difficult to detect from the Earth as nitrogen, and for the same reasons. However, it seemed unlikely that any biosphere had been as copious a source of oxygen as the biosphere on the Earth, and therefore few astronomers thought it likely that oxygen accounted for more than a few per cent of the Martian atmosphere.

The white clouds in the atmosphere were thought, by almost all astronomers, to consist of tiny crystals of water ice, rather like the cirrus clouds in the Earth's atmosphere. Nearly all white clouds were seen to be small, and some areas on the surface seemed to be favoured sites for such clouds. On the Earth, mountain peaks and basins tend to be favoured sites and the same was thought to be the case on Mars. The largest white cloud by far is the sinister-sounding *polar hood*. This is a cloud which grows across the polar regions in each hemisphere during the autumn, thus hiding from view the major phase of seasonal growth of the polar cap. At their maximum extent the polar hoods extend about half-way to the equator. Only in spring does the hood disappear, to reveal a greatly enlarged polar cap. The hoods too were thought to consist of water ice crystals, from which the winter snows fell.

The yellow clouds were widely regarded as clouds of desert dust raised by strong winds. Indeed, sometimes the whole of the Martian globe was seen to be obscured by yellow clouds for several weeks.

Such, in broad outline, was the Earth-based view of Mars in 1965, a view based on observations over distances never less than 55×10^6 km and obtained through a turbid turbulent atmosphere.

On 15 July 1965 Mariner 4 flew by Mars making a closest approach of only 9800 km above the Martian surface. There were two major surprises. First, all twenty-two of the TV pictures that were obtained showed the Martian landscape to be dominated by *impact craters*, the result of bodies colliding at high speeds with the Martian surface. Such craters are abundant on the Moon, and their formation will be outlined in Chapter 6. Their occurrence on Mars indicated lack of geological activity and of extensive weathering by the atmosphere, which would erase such craters. Second, as the spacecraft passed beyond Mars it moved behind the planet as viewed from the Earth. As it did so the radio transmissions from the spacecraft passed through the Martian atmosphere and the changes in the waves induced by the atmosphere enabled the surface pressure of the atmosphere to be determined. And it was not the 100 millibars or so that had been confidently expected but only about 6 millibars. It thus followed that nearly all of the atmosphere could be accounted for by the carbon dioxide that had previously been detected from the Earth.

The dominance of CO_2 led to the revival of an old idea that the polar caps

TABLE 3.1
Successful spacecraft missions to Mars

Date of Martian encounter	Name of mission	Type of mission	Country of origin
14 July 1965	Mariner 4	Fly-by	USA
31 July 1969	Mariner 6	Fly-by	USA
5 August 1969	Mariner 7	Fly-by	USA
13 November 1971	Mariner 9	Orbiter	USA
27 November 1971	Mars 2	Orbiter + lander*	USSR
2 December 1971	Mars 3	Orbiter + lander*	USSR
10 February 1974	Mars 4	Fly-by[†]	USSR
12 February 1974	Mars 5	Orbiter	USSR
12 March 1974	Mars 6	Lander*	USSR
19 June 1976	Viking 1	Orbiter + lander	USA
7 August 1976	Viking 2	Orbiter + lander	USA

* Little or no data returned from the lander.
[†] Intended as an orbiter.

themselves were not made of water but of carbon dioxide, which also forms a white snow. This idea gained support from the next successful missions to Mars, the fly-bys of Mariners 6 and 7 in 1969. The temperature of the south polar cap was measured and found to correspond in the phase diagram of CO_2 to the phase boundary between the solid and the gas at a pressure of a few millibars. This provided strong circumstantial evidence that a polar cap of solid CO_2 was roughly in equilibrium with the CO_2 atmosphere.

On 14 November 1971 the spacecraft Mariner 9 was placed in orbit around Mars. It sent to the Earth pictures and data that were much more extensive and of considerably higher quality than those from the earlier missions. Then, in 1976 the two spacecraft Vikings 1 and 2 went into orbit around Mars and from each of them came a small Lander which were successfully landed on the Martian surface. The pictures and data were even better than from Mariner 9. However, it was with Mariner 9 that after 350 years of Earth-based exploration of Mars we began to learn what its atmosphere and surface were really like: all will be revealed in sections 3.3 and 3.4.

Table 3.1 lists the successful spacecraft missions to Mars. The series of USSR spacecraft between Mariner 9 and Vikings 1 and 2 returned comparatively little information.

3.2 The Martian interior

In principle the same methods are available to study the Martian interior as have been used to study the Earth's interior. In practise, some of these methods have not yet been applied to Mars or have yielded little useful information. Gravitational studies have been of considerable use in relation to both planets and in relation to most of the other planets also. This is an appropriate moment to consider how such gravitational studies are made.

● *3.2.1 Gravitational probing of planetary interiors*

Figure 2.8 (a) in Chapter 2 illustrates the gravity field around a body which has a spherically symmetrical mass distribution, to which the planets closely approximate. The equal length arrows indicate that the field is spherically symmetrical and therefore has a constant strength on a spherical surface centred on the centre of mass of the planet. The field strength diminishes with distance in proportion to the *square* of the distance from the centre of mass. This is called the *inverse-square law*. At a given distance the strength is proportional to the mass of the planet. In the case of the Earth the field strength can be measured at the surface of the Earth and this gives the mass. It can also be calculated from the orbits of satellites, natural (the Moon) or artificial. It can be shown that if the semi-major axis and orbital period are known, and in the case of the Moon, because its mass is not negligible, the position of the centre of mass of the Earth–Moon system, the mass of the Earth can be calculated. These two methods, based respectively on surface and orbital measurements, are applicable to other planets.

The mass and the volume of a planet yield the mean density. But gravity measurements can also yield information about the variation of density with depth. This is possible only if the mass distribution is *not* spherically symmetrical: if it *is* spherically symmetrical then the external gravitational field is also spherically symmetrical and would be the same for all spherically symmetrical mass distributions of given total mass. Fortunately, no planet has a perfectly spherically symmetric mass distribution. The main reason is the axial spin which *flattens* the planet in the manner shown in Figure 3.5 (a). This is measured by the polar flattening *f* defined in section 2.1.3.

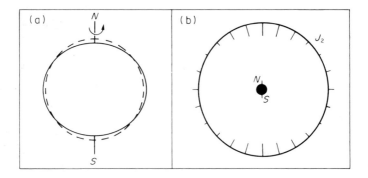

FIGURE 3.5
(a) Deformation of a spinning planet. (b) The J_2 component of the gravitational field.

A particularly important component of the total departure from a spherically symmetrical gravitational field is illustrated by Figure 3.5 (b). The short lines show departures from the spherically symmetrical value, the lines pointing inwards showing where the field strength is less and those pointing outwards showing where it is more the length of the line being proportional to the departure. The characteristics which define this component are that it is the same at all longitudes and that the departure is zero *once* in each quadrant. The strength of the component is measured by a quantity J_2 (= "jay-two"). This is one of many *gravitational coefficients*. Other coefficients measure the strength of components that have more zeroes in each quadrant, and some vary with longitude.

The greater the number of zeroes of a component the more rapidly its strength falls off with distance from the planet. A long way from a planet only the spherically symmetrical field can be sensed. At some intermediate distance the $\mathcal{J}_2$ component can also be sensed. Yet closer the other components become appreciable, though there is no danger of confusion with the $\mathcal{J}_2$ component because each coefficient is associated with a variation of specific spatial form. Note that it is not the value of $\mathcal{J}_2$ (for example) that varies with distance but the gravitational field component with which it is associated, that is, the distance variation is not included in $\mathcal{J}_2$.

For all planets the value of $\mathcal{J}_2$ is larger than the values of the other gravitational coefficients, and in most cases the other coefficients may be neglected when determining the variation of density with depth. For the remainder of this section I shall assume that this is the case.

$\mathcal{J}_2$ can be obtained from gravity-field measurements made across the surface of a planet or from the orbits of satellites, which are not quite the ellipses they would be in a spherically symmetrical field, or from the paths taken by fly-by spacecraft. However, though $\mathcal{J}_2$ does help to constrain models of density variations with depth, a much more powerful constraint is obtained if $\mathcal{J}_2$ is used with other information to calculate the *polar moment of inertia, C.*

The *moment of inertia* of a body is a measure of the readiness with which its rate of rotation may be changed, the greater the moment of inertia the harder it is to change its rotation rate. For a body of given mass and volume the more the mass is dispersed towards the surface the greater the moment of inertia. In Figure 3.6 two extreme cases are shown for a spherical body of mass M and radius R. In Figure 3.6 (a) the mass is in a spherical shell at the surface and in Figure 3.6 (b) it almost entirely resides in the centre. It would be far easier to change the rotation rate of the latter body than of the former body.

The polar axis of a planet, by definition, is the axis around which it spins, and the polar moment of inertia (C) is a measure of the readiness with which its rate of rotation around this axis may be changed. Clearly, as indicated in Figure 3.6, if C can be measured then some indication of the density variation with depth will be obtained. Unfortunately it is not possible to

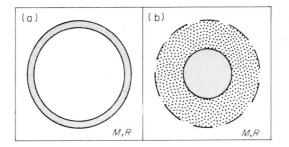

FIGURE 3.6
(a) Mass M is concentrated in a shell of outer radius R. (b) Mass M within the sphere radius R is largely confined near the centre.

directly measure C, because it is not possible to identify with sufficient accuracy the forces causing the *very slight* changes in rotation rate of the planets. Indeed, it is only for the Earth that the changes in rotation rate have themselves been measured.

C thus has to be obtained indirectly. One method involves the moment of inertia around any axis lying in the equatorial plane of a planet which will differ from C if the planet is not spherically symmetrical. Provided that the planet's departure form spherical symmetry is largely a result of its axial spin then the moment of inertia around such an axis is independent of its orientation in the equatorial plane. Let A denote its value. The *ratio* of A to C can be measured through the rate of *polar axis precession* which is illustrated in Figure 3.7: the direction of the polar axis, with respect to the distant stars, very slowly changes. This is a consequence of a non-spherically symmetric mass distribution and its rate can be shown to give the *ratio* of A to C. It can also be shown that J_2 yields the *difference* between A and C. Therefore, both A and C can now be determined. The only other information required is the equatorial radius R_E of the planet, its mass M, and its sidereal axial period T_a.

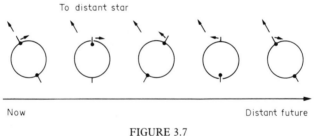

To distant star

Now   Distant future

FIGURE 3.7
The precession of the spin axis of a planet.

Unfortunately, it is only for the Earth that the rate of polar axis precession has been measured. This is partly because of the long series of observations that have been made from the Earth, and partly because the Moon acts on the non-spherically symmetrical mass distribution of the Earth to produce a comparatively rapid rate of precession, thus making the rate easier to determine. The Earth's axis takes 25 725 years to complete the precessional cycle in Figure 3.7. The model of the Earth's interior in Figure 2.3 is consistent with the calculated value of C. In the case of the Moon a quantity as useful as the rate of precession has been measured, as you shall see in section 6.2.1.

For the other planets, in order to calculate C it is necessary to assume that throughout most of their interiors they behave *as if* they were liquids. It is *not* an assumption that the interior of a planet is liquid, but that the interior has adjusted itself to the effects of axial spin and external forces in the same way that a liquid would have done. If this is the case then the interior is in *hydrostatic equilibrium*. Solids can achieve hydrostatic equilibrium in planetary interiors because in a large mass gravity is the dominant internal force and over long periods can induce hydrostatic adjustment. The period required falls as the internal temperatures rise, because of the increasing *plasticity* of the interior.

If a planet is in hydrostatic equilibrium then *hydrostatic theory* applies and it can be shown that C can be calculated provided that M, R_E and any *two* of J_2, T_a and f are known.

Departures from hydrostatic equilibrium can be revealed by the gravitational coefficients *other* than J_2. In some cases such departures are largely confined to the outer regions of a planet because it is cooler there or because of strong surface activity. In this case f differs from the value it would have had, had the planet been in hydrostatic equilibrium. However, J_2 would not differ as much as f from the case of hydrostatic equilibrium because J_2 is associated with the *whole* mass distribution whereas f is particularly sensitive to the *near-surface* regions. Furthermore, the gravitational coefficients other than J_2 can be used to adjust J_2 to a value that can be used with hydrostatic theory. Therefore, when calculating C it is preferable to select J_2 and T_a out of J_2, T_a and f.

Very small-scale departures from hydrostatic equilibrium, even on the scale of mountains, are immaterial on a global scale.

3.2.2 Gravitational probing of the Martian interior

The mass of Mars was unknown until the discovery of Phobos and Deimos in 1877. Following their discovery the mass of Mars was determined from their orbital periods and semi-major axes as outlined in the preceding section. The mass was similarly determined from the orbits of spacecraft, and the most accurate value has been obtained from a combination of spacecraft orbits and spaceraft observations of the orbits of Phobos and Deimos. In terms of the mass of the Earth (M_E) the mass of Mars is $0.1074 M_E$.

The mean density of Mars is then obtained by dividing its mass by its volume and this yields 3940 kg/m^3. The mean density of the Earth is 5520 kg/m^3. However, in making interplanetary comparisons it is necessary to allow for different degrees of internal compression in the different planets. This compression arises from the downward pressure of material on the materials at greater depth which in turn arises from the gravitational attraction of one part of the planet on another, the greater the depth the greater the compression. The degree of compression depends on the materials, and in the case of the Earth, for any materials that are at all likely to be abundant, the effect of this compression is such that were it to be removed the Earth would expand to give a mean uncompressed density of between 4000 and 4500 kg/m^3. In the case of Mars the mean uncompressed density would lie between 3700 and 3800 kg/m^3. This smaller range for Mars is *not* because the internal composition of Mars is better known than for the Earth, the reverse is the case, but because the mass of Mars, and consequently the degree of compression, is considerably less and can therefore be allowed for with greater accuracy.

The uncompressed mean densities of the Earth and Mars are not as different as the compressed mean densities. However, the uncompressed density of Mars is clearly less than that of the Earth and this indicates a difference in composition.

The value of J_2 has long been known from the orbits of Phobos and Deimos, though spacecraft orbits have yielded far more accurate values and have also yielded other gravitational coefficients which indicate that Mars is not quite in hydrostatic equilibrium. There is evidence that much of the departure is associated with a surface feature called the *Syria Rise* (see section 3.3). The polar moment of inertia (C) has been calculated from J_2 and T_a as outlined in the preceding section, and some scientists have thought it necessary to apply small corrections to J_2 in order to apply the hydrostatic theory which yields C, whilst others have used the observed values of J_2.

In all cases the values of C show that the density of Mars increases towards the centre more than can be explained by the increasing compression of a homogeneous body as the centre is approached: the materials nearer the centre have a greater uncompressed density than those at shallower depths. However, *if* Mars has an iron core and a rocky mantle then the core must occupy a smaller fraction of the volume of Mars than the fraction of the Earth's volume occupied by its iron core, otherwise the value of C for Mars would be smaller than that deduced from observations.

3.2.3 Other probes of the Martian interior

Much of what we know about the Earth's interior comes from seismic waves. Both the Viking Landers carried seismometers, but only the one on Viking 2 worked. During the daylight hours at this site on the Martian surface the winds were sufficiently strong to mask any seismic waves. At night the winds were less, but even so over a period of several years only one or two signals were detected that could have come from "Marsquakes". This indicates that the Martian interior is more quiescent than that of the Earth.

No Martian magnetic field of internal origin has been detected. (Weak fields associated with the ionosphere and surface rocks have been.) This means that the dipole moment (see section 2.1.7) is less than $1/4000$ of that of the Earth. If the dynamo theory is right then either Mars has no liquid conductors inside it or they are of extremely small volume. Therefore, any liquid iron core must be extremely small or non-existent. This does not rule out a *solid* iron core.

No measurements have been made of heat flow from the Martian interior.

3.2.4 Models of the Martian interior

Figure 3.8 shows three of many models of the Martian interior. All three are consistent with the various constraints outlined above and also with the solar relative abundances of the elements and the volatility of such elements and the compounds they are likely to form. The variety of plausible models is greater than for the Earth because the observational constraints on the Martian interior are fewer and weaker. However, there is a consistent and clear difference between models of the Earth and of Mars, namely, that whereas about 15% of the Earth's volume resides in a predominantly liquid iron core, any iron core in Mars accounts for a smaller fraction of the volume of Mars, and much, perhaps all of such iron is compounded with sulphur to form iron sulphide (FeS). Moreover, the small upper limit of the magnetic dipole moment indicates that any such core is likely to be largely solid.

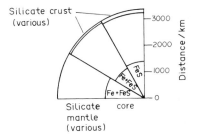

FIGURE 3.8
Three models of the Martian interior.

The solid core and the low level of seismic and geological activity (section 3.3) indicates that temperatures are lower inside Mars than inside the Earth. This comes as no surprise. Mars is smaller than the Earth, and therefore, per unit mass, can lose energy more rapidly to space. Thus, if the Martian internal energy sources, per unit mass, are comparable to or less than those in the Earth then lower internal temperatures are expected today and throughout much of Martian history.

3.3 The surface and near-surface regions of Mars

I shall discuss the topographic features first, and then the albedo features. However, it is first necessary to outline how the altitude of a point on a planet's surface is defined.

● *3.3.1 Altitude*

If a planet has a perfectly spherically symmetrical distribution of mass then its surface will be spherical and the centre of mass will be at the centre of the sphere. In this case the *altitude* would be measured with respect to the centre of mass and the surface everywhere has the same altitude. An obvious choice for the zero of altitude is this spherical surface. If such a planet now spins and reaches hydrostatic equilibrium then the centre of mass remains in the same place but the surface is no longer spherical, as shown in Figure 3.5 (a). If altitudes were still measured with respect to a spherical surface centred on the centre of mass then clearly the equatorial regions would be higher than the polar regions. However, to someone inhabiting the surface of the planet the poles would not appear to be "downhill". For example, water placed anywhere on the surface will not tend to flow towards the pole (nor towards the equator). This is because the surface is one of hydrostatic equilibrium. It is therefore natural to use the surface of hydrostatic equilibrium as the "zero" of altitude.

No planet is exactly in hydrostatic equilibrium, particularly on a small scale, and therefore the surface is not all at zero altitude. A reference surface of zero altitude thus needs to be chosen. The surface of a liquid is a surface in hydrostatic equilibrium, and therefore in the case of the Earth *sea-level* is chosen as the zero of altitude. Any given pressure in an atmosphere is also a surface in hydrostatic equilibrium. Therefore, in the case of Mars, which has no oceans, the zero of altitude is chosen where atmospheric pressure is 6.00 millibars. Actual surface pressures vary from about 2 to about 12 millibars.

For any planet a spherical surface can be selected which provides the closest fit to the actual surface of the planet. The centre of this surface is called the *centre of figure* as illustrated in Figure 3.9. It need *not* correspond to

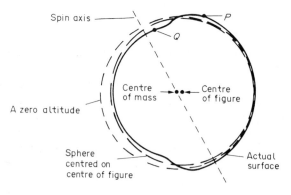

FIGURE 3.9
The centre of mass and the centre of figure of a planet.

the centre of mass. In Figure 3.9 the centre of figure is offset to the right of the centre of mass because of the greater prevalence of lower density materials to the right. A surface of zero altitude, obtained as outlined above, is also shown and you can see that the actual surface to the right lies at a higher altitude than that which lies to the left, though the mean altitude difference need not equal the offset.

3.3.2 Martian surface topography

On Mars the altitude of much of the surface has been obtained from *spacecraft occultations*, that is, the passage of a spacecraft behind Mars as detected from the Earth by the abrupt cessation of radio signals from the spacecraft until it emerges again. Measurements made just preceding the occultation yield the atmospheric pressure at the point of the surface at which the occultation occurs, from which the altitude can be deduced. On a small spatial scale altitudes were deduced from the time lapse between sending down a *laser* pulse from a spacecraft and receiving the reflection from the surface: this is similar to a technique used on Venus using *radar* and will be outlined in section 4.3. Both types of techniques require that the *position of the spacecraft* be accurately known: section 4.1.1 outlines how this is accomplished.

Figure 3.10 is a map of Mars showing topographic and albedo features. It also shows a curved line which on a globe would correspond to a line dividing the Martian surface into roughly two hemispheres. To the south of this line the altitudes are on average higher than to the north. The dividing line is inclined with respect to the equator and I shall therefore refer to the south*erly* and north*erly* hemispheres rather than to the southern and northern hemispheres. Detailed measurements show that though both hemispheres, and particularly the northerly hemisphere, display considerable altitude variations, the mean altitude of the southerly hemisphere is about 3 km greater than the mean altitude of the northerly hemisphere. Therefore, as indicated in Figure 3.9, the centre of figure must be offset from the centre of mass roughly in the direction of the southerly hemisphere. The amount of offest is 2.5 km.

One way of accounting for the altitude difference is by means of a crust of lower density than the mantle beneath and of variable thickness such that the crust is on average thicker in the southerly hemisphere than in the northerly hemisphere. However, many other hemispherical differences in the variation of density versus depth would also account for the altitude difference.

The next largest topographical feature on Mars has been unofficially named the *Syria Rise*. This is a great dome which in Figure 3.10 lies roughly between longitudes 40° and 140° and between latitudes +40° and −40°. Measurements of the gravitational field of Mars have revealed that the Syria

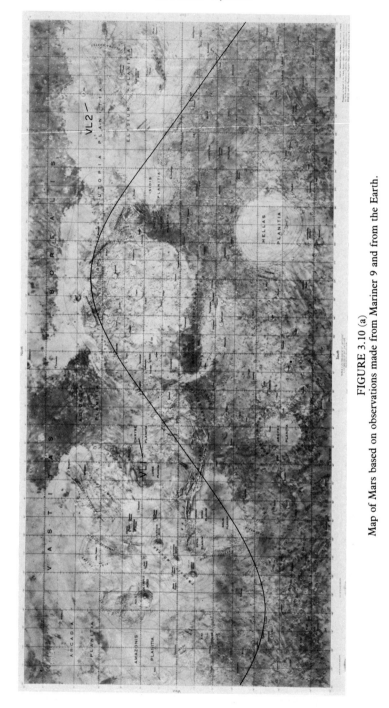

FIGURE 3.10 (a)

Map of Mars based on observations made from Mariner 9 and from the Earth.

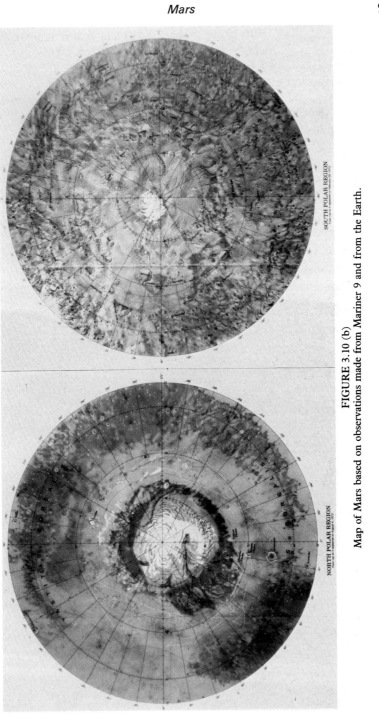

FIGURE 3.10 (b)

Map of Mars based on observations made from Mariner 9 and from the Earth.

Rise is *not* isostatic equilibrium (section 2.1.10) and that it represents the furthest and most extensive departure from isostatic equilibrium on the planet. To acquire isostatic equilibrium its altitude would have to be reduced. This indicates that it may be being borne aloft by interior convection. However, the seismic quietness of Mars is evidence against this possibility. An alternative explanation is that the lithosphere is thick thus requiring enormous times to adjust itself to isostatic equilibrium. You will see in section 3.3.5 that the Syria Rise is a relatively recent Martian feature and therefore regardless of how it was produced, and this is unknown, a thick lithosphere could well explain its present lack of isostatic adjustment.

The Syria Rise is scarred by great cracks, of which the most spectacular is *Valles Marineris*. This can be seen in Figure 3.10, and a portion of it is shown in detail in Figure 3.11. These cracks show that there was uplift of the lithosphere, resulting in it cracking under the tension produced. The largest

FIGURE 3.11
A portion of the main branch of Valles Marineris on Mars, imaged by Viking Orbiter 1. This part of the main branch is about 75 km wide.

lithospheric crack that currently scars the surface of the Earth is the Rift Valley in Africa, but in spite of the Earth's larger surface area Valles Marineris is larger. However, it seems that many of the Martian cracks, including Valles Marineris, have been enlarged, perhaps by wind erosion or by removal of magma.

Mars also bears *volcanoes* that are larger than any known on the Earth. The least eroded, and therefore the youngest, are on the western half of the Syria Rise, an area named the *Tharsis Ridge*. There are four volcanoes there, visible in Figure 3.10, all of which rise to altitudes of about 27 km, making them the highest points on Mars. Their volcanic nature is apparent from the details of their topographic form. They do not seem to be currently active. Figure 3.12 shows *Olympus Mons*, the largest of all.

The closest analogues on Earth are *shield volcanoes*, particularly those that occur *within* plates rather than at plate boundaries. They derive their name from their profile: broad domes resembling a warrior's shield viewed edge on. On Earth they are produced by lavas rich in basalt (= *basaltic lavas*) and this interpretation applied to Mars is consistent with spectroscopic studies from orbit and with observations and analyses made at the surface. It could thus be that the lavas which built these volcanoes are the result of a partial melting of a mantle broadly similar in composition to that of the Earth. The interior models in Figure 3.8 are in accord with this.

The great size of the Martian volcanoes could be the result of a copious source of lava plus lack of local isostatic adjustment plus lack of lithospheric plate motion which would tend to produce, as on the Earth, a chain of smaller volcanoes. The last two points support the view that the lithosphere is thick.

There are other volcanoes on Mars of various sizes and in various states of preservation plus indications of numerous other types of volcanic features from some of which copious amounts of lava may have flowed.

But though Mars may have experienced an appreciable amount of volcanic reworking of its surface, perhaps producing a widespread crust, it seems clear that no system of Plate Tectonics ever developed. This conclusion is strongly supported by the clear absence from Mars, see Figure 3.13, of topographic features characteristic of plate boundaries, such as mountain ranges, mid-"ocean" ridges, "ocean" trenches and volcanoes of a type found on Earth at plate boundaries and which are unlike shield volcanoes. This conclusion is also strongly supported by the hysometric distribution of Mars. This is also shown in Figure 3.13: altitudes on Mars show little tendency to cluster around two levels, unlike the Earth (Figure 2.9 (b)) where the two levels correspond to oceanic crust and continental crust and which are generated by Plate Tectonics. On Mars there is no evidence for the existence of two types of crust.

Perhaps the nearest that Mars has come to Plate Tectonics is the lifting of the Syria Rise, representing the beginnings of plate creation and motion.

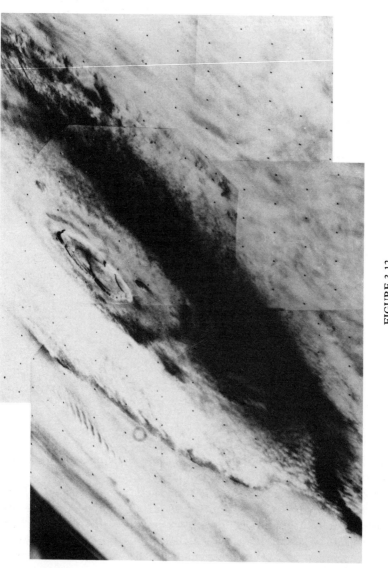

FIGURE 3.12

Olympus Mons on Mars with cloud lying high up its flanks, imaged obliquely by Viking Orbiter 1. The multi-ringed cauldera on the summit is about 80 km across.

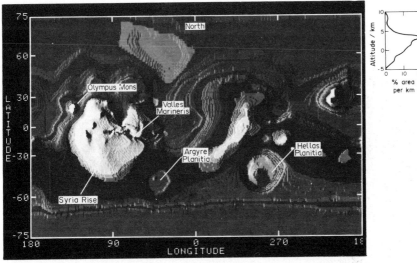

FIGURE 3.13

A drawing of Mars emphasizing altitudes, and the corresponding hypsometric distribution.

Impact craters are fairly common on Mars. However, they are considerably more common in the southerly hemisphere than in the northerly hemisphere, and this is another marked difference between the two hemispheres. Figure 3.14 shows a region of Mars which happens to lie in the southerly hemisphere but in which the less cratered area at middle left is fairly typical of the cratering in the northerly hemisphere.

The formation of impact craters is outlined in section 6.3.1 and in section 3.3.4 their use in establishing the history of a planet is described. In section 3.3.5 I shall return again to Martian craters and to the hemispherical difference in their abundance.

3.3.3 Martian albedo features

Figure 3.10 also shows the Martian albedo features. In broad outline many of these are visible from the Earth.

Within most of the dark areas the terrain is streaky, as illustrated in Figure 3.15. This strongly suggests wind-blown dust, and ample evidence that Mars is a dusty place comes from the yellow clouds which are clearly dust storms, and from the Landers which clearly reveal dust at the Martian surface as shown in Figure 3.16. These Landers have also detected strong winds which would move the dust.

It has been established that the bright material is *fine* dust and that it is considerably more mobile in the winds than the dark material. In most cases the dark material underlies the bright material and is too coarse to be moved

FIGURE 3.14
Argyre Planitia (middle left) and more heavily cratered Martian terrain surrounding it, imaged obliquely by Viking Orbiter 2. Argyre Planitia is about 900 km
across.

by the winds. The main exception is the dark collars which encircle the polar
regions (Figure 3.10). These collars consist of dust dunes that have
presumably been deposited by the wind. Elsewhere, most of the *dark* albedo
features consist of regions where a good deal of dark material is exposed. By
contrast, the *bright* albedo features are regions where the bright dust is far
more predominant.

The yellow clouds consist largely of bright material.

There is considerable evidence that the bright material is derived from the
dark material by a variety of physical and chemical processes that lead to the

FIGURE 3.15
Wind-blown bright dust on Mars, imaged by Viking Orbiter 2. The largest crater
is about 5 km across.

formation of a fine bright red dust. There are several types of dark material from which is mixed a bright dust of fairly uniform composition across Mars. All of the various types of dark material are also red, the red tint of bright and dark materials being caused by certain iron compounds. In the case of the bright material it is caused by an oxide of iron called ferric oxide and this could be the result of the combination of oxygen with iron compounds in the dark materials, the oxygen being taken from the atmosphere. The ferric oxide may be confined to a surface layer on the dust grains.

The bright red deserts of Mars are thus red for much the same reason that common rust on Earth is red.

It is clear from Figure 3.10 that there is no correlation between the albedo features and the large-scale topographic features. However, there is a correlation between the albedo features and small-scale topography plus the strength and direction of the winds, though this correlation is intricate and not fully understood.

The seasonal variations in the albedo features are the result of the seasonal variations in the winds, and the non-seasonal variations are largely the result of longer-term shifts in the wind pattern. The variations in the features arise from changes in the total area of dark material exposed through the combined effects of wind scouring and deposition of bright dust from dust storms. The deposited dust can entirely obscure dark materials in some places and in other places can deposit a thin veneer which slightly brightens the terrain.

Thus, in place of dark-green vegetation waving in the Martian winds there

FIGURE 3.16
A view of the Martian surface, imaged by Viking Lander 1. The horizon is about 3
km away. The larger boulders are about 1 metre across. The spacecraft, rather
than the horizon, is tilted.

are dark red materials, and in place of the seasonal growth and decay of plants
there are seasonal variations in the winds and their effect on the mobile bright
dust. There is no sign of a correlation between darkening and local spring nor
of any wave of darkening moving equatorwards in the spring hemisphere.

● *3.3.4 Craters as chronometers*

Craters can be used to date surfaces, and this is of particular importance
when radiometric dating has not been done. Only terrestrial and lunar
samples have been radiometrically dated.

If impact craters are not removed then the longer a surface is exposed to
bombardment the greater the number of impact craters it bears, until it is
saturated. On a saturated surface each new impact will *on average* obliterate
one previously existing crater and therefore there is no net change in the
number of craters on the surface. Of course, a large impact will create a large
crater and may obliterate several smaller craters but a small impact may

obliterate no craters at all. The one-for-one swop is an average outcome.

Many planetary (and satellite) surfaces have regions that are saturated with impact craters and other regions that are unsaturated. In the likelihood that the whole surface was roughly equally bombarded it follows that the less cratered the surface the more recently it has been reworked. Thus, different regions of a surface can be placed in a time-order.

The clearest event which can be placed in a time-order is a short-lived complete resurfacing which removes all previous craters. Sometimes such removal will be incomplete, especially of the larger craters the rims of which may still be visible. In this case the fresher less degraded craters, some of which may cut through the older more degraded craters, enable the *partial* resurfacing to be time-ordered. The degraded remnants of the older craters enable the original surface to be time-ordered provided that allowance is made for the greater fraction of the smaller craters that will have been obliterated than of the larger ones.

If there is a gradual obliteration of craters then there is clearly no short-lived event to be time-ordered. The present number of craters indicates the accumulated effect of this gradual obliteration.

The state of degradation of craters can itself help place the various regions of a surface in a time-order, the greater the degradation the longer the time for which the surface has been exposed to the degrading processes.

The degradation and obliteration of craters can be caused by a variety of means. A region may melt, there may be copious flows of lava or of dust, there may be erosion through impacts too small to produce craters, there may be seismic waves, and there may be weathering by the atmosphere.

The *absolute* ages of cratered regions have only been obtained for the surfaces of the Earth and the Moon, by radiometric dating. Because of continuing intense resurfacing of the Earth, impact craters are scarce and are almost all comparatively young. Therefore, the absolute ages of the various regions of the Earth's surface provide a cratering rate only for comparatively recent times. However, for the Moon a rough graph of crater production rate throughout the last 3900 Ma has been obtained and will be outlined in section 6.3.5. But, as you shall see in Chapter 8, it is not clear how this graph may be applied to other bodies, and therefore a wide range of absolute ages of cratered regions can be deduced for such bodies depending on various assumptions that have to be made in applying the lunar graph to them.

3.3.5 *Craters, and the evolution of the Martian surface*

There is a great range of crater sizes on Mars as on all significantly cratered bodies extending from basins such as Hellas Planitia (Figure 3.10) down to craters a few kilometres diameter. Smaller craters are scarce compared to the Moon and this is clearly because the Martian atmosphere would weather away small craters relatively rapidly. This is also the case for other topographically low features, such as the *ejecta blanket* which is material thrown out by an impact and which extends from the crater rim.

Number densities (number per unit area) range from saturation in some small regions of the southerly hemisphere to an almost complete absence on the flanks of the four Tharsis Ridge shield volcanoes indicating the relatively recent origin of these volcanoes. Craters are also scarce in the polar regions except where pits have been eroded revealing a more heavily cratered terrain beneath several hundred metres of dust showing that in the polar regions craters have been buried beneath dust. Otherwise the polar regions bear a few partially buried large craters and a handful of small craters which must post-date the dust. The predominance of small craters on the polar deposits results from the greater abundance in space of smaller than of larger impacting bodies.

Deposition and removal of dust has also resulted in the resurfacing of the Martian surface elsewhere. There has also been some resurfacing and degradation through the removal of subsurface volatiles: more on this in section 3.4.

The northerly hemisphere is less heavily cratered than the southerly hemisphere. Widespread resurfacing of the northerly hemisphere has thus occurred and the fashionable explanation is that it has been subjected to more copious flows of lava than the southerly hemisphere. Dust blanketing may also be important though most scientists believe that except in the polar regions the bright dust is a thin veneer. There is evidence for layering in the walls of Valles Marineris and elsewhere and most scientists believe this to be lava sheets, though it could be layers of dust.

There are numerous features in the northerly hemisphere suggestive of lava sources, such as small domes and cones which could be small volcanoes. Indeed, some of them are so unweathered that they could be very recent and some may even be dormant rather than dead. There are also features which may be fissures from which lava oozed and broad channels along which lava may have flowed. Moreover, many sources of lava of low profile may be buried beneath the lava to which they gave rise.

The southerly hemisphere also contains features suggestive of lava, from heavily weathered almost flat shield volcanoes to lava channels. However, there are no fresh features to suggest relatively recent activity, and the long-ago activity seems not to have been as widespread as in the northerly hemisphere.

If there has been this difference in volcanic activity between the two hemispheres then it could be a result of a thinner and therefore weaker crust in the northerly hemisphere other evidence for which was outlined in section 3.3.2. However, it is hard to see how such a variable thickness crust could have been produced.

Absolute ages of various regions of the Martian surface have been estimated by applying the lunar graph of cratering rate with suitable modifications to Mars. It has been estimated that most of the southerly hemisphere is over 3000 Ma old and that the small regions saturated with craters may be over 3900 Ma old. The least cratered regions may be no older than a few hundred Ma, and the polar regions may be even younger. The ages of the "young" Tharsis Ridge volcanoes are very uncertain largely because of the rather small areas they cover. Different parts of their flanks have different number densities of craters indicating eruptions spread over a long period. Some estimates place the origin of these volcanoes about 2500 Ma ago, other estimates are about 1000 Ma. There is also disagreement about when the volcanic activity ceased, if indeed it has.

Roughly 50% of the Martian surface is probably older than 3000 Ma, and roughly 85% is probably older than 2500 Ma. The corresponding figures for the Earth are nearly zero and about 1% respectively: the Earth has been and continues to be far more geologically active than Mars. As outlined in section 3.2.4 the Martian interior may well have been cooler than the Earth's interior throughout much of Martian history and therefore this lower level of geological activity is to be expected. Moreover, there are some reasons to believe that the interior of Mars today is cooling. Therefore, geological activity on Mars is very probably in decline. Plate Tectonics never got going, and it seems unlikely that it ever will.

3.4 Atmosphere, surface volatiles and life

3.4.1 The Martian atmosphere today

The composition of the Martian atmosphere, in terms of relative numbers of molecules, is as follows: 95% CO_2; 2.7% N_2; 1.6% Ar; 0.13% O_2; 0.07% CO; about 0.03% H_2O (lower atmosphere); plus traces of several other gases. The atmosphere of Mars accounts for a fraction of the total mass of the planet that is about 24 times less than the fraction of the Earth's mass that is accounted for by its atmosphere. Water may seem to be scarce in the Martian atmosphere but the above figure is about the maximum that the atmosphere can hold at the low surface temperatures of Mars, about 218 K being the global average. It is therefore possible that a good deal of water exists in solid form at and below the Martian surface. The oxygen in the Martian atmosphere is largely the result of the photodissociation of CO_2 by solar uv radiation. This yields CO and O, with O_2 forming subsequently. Very little ozone forms because of the low O_2 abundance and therefore a good deal of solar uv radiation penetrates to the surface. Water is also photodissociated, to yield OH and H. Further photodissociation of OH yields small quantities of O_2 but a more important consequence is that OH via a complicated sequence of chemical reactions causes CO and O to recombine at a greater rate than would otherwise be the case, thus reducing the amount of CO_2 that is dissociated at any one time. This reduces the oxygen abundance, and it is further reduced through combination with surface materials, one consequence of which is the conversion of dark-red material to bright-red material.

The vertical structure of the atmosphere, averaged over the surface and over the seasons, is shown in Figure 3.17. Typical altitude ranges are shown for the various types of clouds. Water clouds consist of tiny ice crystals and can occur down to ground-level where they form icy ground fogs. The polar hood consists largely of tiny ice crystals of CO_2 rather than H_2O, and smaller CO_2 clouds can occur elsewhere.

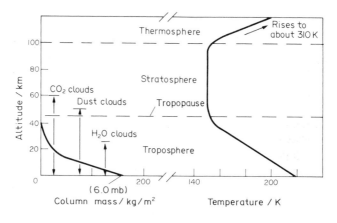

FIGURE 3.17
The vertical structure of the Martian atmosphere.

The variation of temperature with height divides the Martian atmosphere into layers. Like the Earth the lowest layer is called the troposphere in which the temperature decreases with height. However, there is usually sufficient dust in the Martian troposphere for direct absorption of solar radiation to make the lapse-rate smaller than the adiabatic value. The dustiness of the troposphere is usually sufficient to make the sky appear yellow or orange from the Martian surface. Dusty skies on Earth can also appear the same colours. The more usual blue colour on Earth is the result of the scattering of light by molecules in the atmosphere, which scatter blue light more efficiently than they scatter red light. A dust-free atmosphere on Mars would appear a very deep blue.

The Martian thermosphere is much cooler than the Earth's thermosphere because CO_2 radiates energy to space more efficiently than the common molecules in the Earth's thermosphere. Thus, the solar uv radiation absorbed in the Martian thermosphere does not cause a large temperature rise as it does on the Earth.

High above the troposphere lies the Martian ionosphere. The dominant ions O^+ and O_2^+ as in the Earth's ionosphere though those on Mars come mainly from CO_2 whereas those on Earth come mainly from O_2.

Diurnal variations are fairly extreme on Mars. For example, at both the Viking Lander sites (Figure 3.10) in local summer the diurnal variation 1.6 m above ground level was from about 190 K to about 240 K. At ground level the temperatures are estimated to be about 30 K higher but with a comparably large range. Seasonal variations are also extreme. These extreme diurnal and seasonal variations are partly because there are no oceans, which on Earth act as large and accessible heat reservoirs, and partly because of the low mass of the Martian atmosphere.

The low mass of the atmosphere also results in a rather weak greenhouse effect which causes a rise of only about 5 K in the global mean surface temperature, and it also gives a weak poleward transport of solar energy from the equator where the greatest amount of solar energy is absorbed as on the Earth.

Precipitation from the atmosphere occurs in several forms: dust; small crystals of water ice; small crystals of CO_2 ice; and CO_2 *clathrates* in which CO_2 molecules are "imprisoned" in the rather open cage-like structure of water crystals. These various forms of precipitation of ice crystals I shall call *whitefall*. In all cases the whitefall would be a fine white dust.

The most dramatic whitefall occurs in the polar regions during autumn and winter. The whitefall at first consists of water ice, then of clathrates, and then of CO_2 ice as the temperature drops. In the spring and summer the caps shrink by evaporation.

3.4.2 Surface volatiles today

The polar caps are the most obvious repositories of surface volatiles. Each consists of a *seasonal cap* which grows to mid latitudes in the winter, and a *residual cap* which is the remnant left in the summer. It is clear that most of the seasonal caps are CO_2: it gets cold enough in winter for CO_2 to condense, and CO_2 is the only sufficiently abundant atmospheric constituent to account for the volume of material in the seasonal caps.

The retreat of the seasonal cap in one hemisphere is not always in balance with the growth of the seasonal cap in the other hemisphere and therefore the amount of CO_2 in the atmosphere oscillates during the Martian orbital period. These cyclic changes have been detected at the Viking Lander sites as a swing in atmospheric pressure of about 15% above and below the local average surface pressures. Each seasonal cap at maximum extent accounts for about 10% of the CO_2 in the atmosphere.

At surface pressures typical of Mars CO_2 condenses at temperatures below about 150 K, and indeed the temperature of the seasonal caps throughout the winter is about 150 K. The temperatures do not fall much below the threshold for condensation because of the *latent heat* released by the CO_2 when it condenses. Latent heat is the heat given out by a substance when it passes from a less dense phase to a more dense phase, and is a direct consequence of a reduction in the average separation of the atoms or molecules in the substance. The heat released tends to "pin" the temperature at the threshold value for condensation.

The *residual* cap at the north pole does *not* seem to consist of CO_2. Observations by Viking Orbiter 2 indicated that the summer surface temperature of the residual cap was about 240 K. This is far too high for CO_2 to exist in condensed form at pressures of a few millibars and therefore the remnant cap must largely consist of the only other plausible candidate for appreciable amounts of whitefall — water. Small quantities of CO_2 could be present, because evaporation is not instantaneous on raising the temperature: latent heat must now be *acquired* by the condensed substance for it to evaporate, and this takes time.

Observations of the *south* polar residual cap indicate that, unlike the north residual cap it may consist of CO_2 presumably with water beneath.

In Figure 3.10 the polar caps are not much larger than the residuals. The spiral patterns are thought to be the result of weathering through the combined effects of wind and Sun.Each arm of a spiral is roughly valley-shaped, and close scrutiny has resulted in cross-sectional models typified by Figure 3.18. The residual cap largely sits on the high terrain between the valleys, and its albedo is rather lower than for clean ice grains suggesting that some rocky dust is present. The deposits beneath, revealed in the valley walls, have a far lower albedo and carry a red tint: they contain a greater proportion of dust.

It is plausible that dust is brought to the polar regions by the same poleward-moving winds that feed the polar hood and the seasonal caps. It would be a minor constituent of the precipitation but could become a major constituent of the surface deposits because of the loss of most of the ices through summer evaporation.

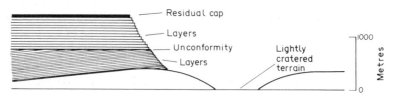

FIGURE 3.18
A simplified cross-section of a typical Martian polar region. The vertical scale is exaggerated.

The *terracing* of the left wall in Figure 3.18, that is, the staircase profile, indicates that these deposits were laid down in layers about 10 to 50 metres thick, this being the step height. The staircase profile could be the result of different proportions of ice and dust in each layer or the result of layer-to-layer variations in the nature of the dust.

Beneath the terraced deposits lie unterraced deposits of bright-red dust with a small amount of ice. Here and there pits have been eroded in this material, as outlined earlier, displaying craters beneath in a reasonable state of preservation. The *un*terraced deposits are several hundred metres thick near the poles. They extend further from the poles than the terraced deposits, gradually getting thinner and perhaps depleted in ices as the polar distance increases. Within about 40° of the equator the bright dust layer is thought by most scientists to be thin in most regions. The dark polar collars (Figure 3.10) consist of dunes of rather coarse dark-red dust.

It is not known whether today there is net erosion or net deposition of material in the polar regions. Nor is it known from where the rocky dust came nor to where it goes. However, it could come from Valles Marineris and similar canyons and from other features outlined below.

Estimates have been made of the quantities of volatiles in the polar deposits that lie above the underlying cratered terrain. Except for the seasonal caps it is thought by most scientists to be unlikely that much CO_2 is locked up in these deposits. However, it is *very* likely that the amount of water in the polar deposits considerably exceeds the amount in the atmosphere. The amount of water in the Martian atmosphere today is between about 0.01 and 0.1 kg/m^2. One estimate is that the water locked up in the polar deposits is equivalent to about 0.25 kg/m^2, but this is very uncertain.

There is considerable evidence that the polar deposits are not the only surface repositories of volatiles on Mars. Figure 3.19 (a) shows a comparatively recent crater around which the ejecta blanket still survives. This blanket strongly suggests that material beneath the surface was liquefied by the impact and spread out at and below the surface rather than being tossed through the air as would happen to molten rock. No such blankets are seen on other cratered planets. Detailed considerations suggest that icy materials were liquefied.

Figure 3.19 (b) shows a typical *outflow channel*. Such channels are probably the result of the sudden melting of ground water which rushes out, leaving behind hummocky land called *chaotic terrain*. Some of the features carved by the water may have been produced by ice, and indeed ice would rapidly form on the surface of any liquid water released on the Martian surface today. In present conditions the ice would subsequently evaporate, except in polar regions.

Figure 3.19 (c) shows some *dendritic* (= "tree-like") *channels*. When channels like these were discovered by Mariner 9 they caused considerable

The Solar System

(a)   (c)

(b)

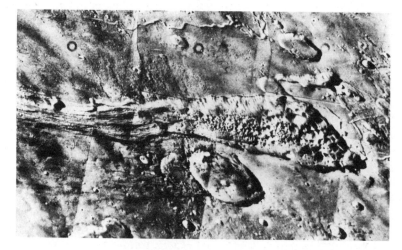

FIGURE 3.19
(a) The Martian crater Arandas, imaged obliquely by Viking Orbiter 1 and
displaying an unusual ejecta blanket. The crater is about 28 km across. (b) An
outflow channel, Capri Chasma, on Mars, imaged obliquely by Viking Orbiter 1.
The width of the picture is about 300 km. (c) Dendritic channels on Mars, imaged
by Viking Orbiter 1. This picture covers an area which measures about 230 km by
150 km.

excitement because they resemble terrestrial river-drainage systems of water deposited by rainfall. The surface pressures on Mars today *are* high enough to permit rainfall, but the temperatures are too low: a thin mist of snowflakes is all that can be expected. Therefore, it seemed probable that Mars was once considerably warmer. However, closer scrutiny of the dendritic channels has revealed several differences from river-drainage systems, and it is now believed that like the outflow channels they are also the result of the melting of ground water though at a much *slower* rate than that which produced outflow channels. The frozen ground water need *not* have been placed there by rainfall.

The heat required to release the frozen water could have been the result of impacts or of geological activity or of climatic warming or of all three. I shall return to the possibility of climatic warming shortly.

The volume of the outflow and dendritic channels is about 1000 times the volume of the water currently existing in the Martian atmosphere and though not all of the channel volume would have been water they do represent a significant repository.

The boundary between the southerly and northerly hemispheres over much of its length has a hummocky appearance rather like chaotic terrain. It has consequently been suggested that the northerly hemisphere lies at a lower altitude than the southerly hemisphere because of the removal of subsurface ices, the hummocky terrain at the boundary resulting from partial removal. An alternative suggestion is that the hummocky terrain is the result of weathering of a once more abrupt boundary, and that the eroded material has made a major contribution to the polar dust deposits. It has also been suggested that the hummocky terrain is the result of the removal of underground magma. Such magma could also account for some of the volume of the chaotic terrain and of the outflow and dendritic channels. Moreover, the heat for the magma could have unfrozen the ground water.

Further evidence for surface repositories of volatiles is provided by the Viking Lander analyses of surface samples. These have been confined to the bright-red dust. The analyses reveal a clay-like composition. Clays, as outlined in section 2.1.9, contain water in hydroxyl form and as water of hydration. The analyses also indicate that some water molecules may adhere to the surfaces of the dust grains, an example of adsorption. In all, about 1% of the dust was water in some form or other. This result had been anticipated several years earlier by Earth-based infrared spectrometry.

The Landers also found carbon dioxide in the dust, mostly absorbed on to grain surfaces though a few percent of the grain mass could be carbonates.

The Landers also measured the atmospheric abundances of the inert gases (section 5.1) and various isotopes of the more common elements particularly N, C and O. From these abundances it can be inferred that substantial repositories of ices probably exist near the Martian surface. There may be the equivalent of no less than about 13 kg/m^2 of water in such repositories though the amount of CO_2 is probably considerably less. Note that these repositories are of water and CO_2 which could readily be exchanged with the atmosphere. Less accessible repositories are excluded, such as those which are very deep, or those in which the molecular binding of water or CO_2 is too strong, as is the case in some of the water that goes to form clays and the CO_2 that forms carbonates. Any nitrates would also contain nitrogen too strongly bound to be counted as readily accessible. Thus, early in Martian history there could have been a more massive atmosphere with much of the volatiles subsequently moving into various surface repositories.

There has also been a gradual loss of volatiles to space. Atoms and molecules of *nitrogen* cannot escape unassisted from the Martian exosphere at an appreciable rate today because the exospheric temperatures are too low. However, there is a series of chemical reactions starting with the dissociation of N_2 into N by solar uv radiation by means of which nitrogen atoms can be accelerated to speeds considerably in excess of their average speed corresponding to the exospheric temperature. The escape rate is correspondingly increased. This is another example of chemical escape (section 2.2.2). It has been estimated that Mars has lost to space about *ten* times as much nitrogen as presently exists in the atmosphere. A similar loss mechanism operates on water and on CO_2. It has been estimated that the equivalent of about 2.5 kg/m^2 of water has been lost by Mars to space though the amount of CO_2 lost is considerably less.

There is thus considerable evidence, much though not all of which has been outlined above, that substantial and accessible repositories of volatiles exist on Mars today and that in the past Mars possessed an even greater endowment. However, it is only possible at present to estimate *lower* limits of the amounts of volatiles remaining today in such repositories. For example, it is not certain whether the fine bright dust covers a thin rubble of darker material which becomes at a shallow depth volatile-free solid rock with few crevices, or whether a loose aggregate of dust and coarser material plus ices and other volatiles extends to a considerable depth. Nor is it possible to tell with certainty whether volatiles on Mars make up a larger or smaller fraction of the mass of the planet than is the case on the Earth: more on this in Chapter 5.

3.4.3 Atmospheric and climatic changes

On the most heavily cratered and therefore oldest terrain only large craters have survived weathering. Neverthless they bear the scars of weathering so intense that had it persisted into the lifetime of small craters on slightly less heavily cratered and therefore slightly younger terrain these small craters would have been completely eradicated. It therefore seems likely that the rate of weathering declined in the early history of Mars. This suggests a fall in atmospheric mass.

In order to produce the outflow and dendritic channels it is not essential for atmospheric conditions to have been different in the past. Ground water could have been melted by impacts and by geological activity, and though liquid water is not a stable phase on Mars today it could last long enough beneath a protective cap of ice to yield those features in the channels which seem to have been carved by liquid water. Most of the outflow channels lie in the equatorial zone and though this could indicate the role of tropical warmth they also lie near the boundary between the northerly and southerly hemispheres and thus could arise from a geological cause. However, the dendritic channels are not clearly linked with likely centres of geological activity and bear few signs of ice-carved features. They may therefore have been carved by relatively unprotected liquid water and this is easier to understand if Mars was then warmer than it is today.

The polar deposits suggest an intricate series of atmospheric changes. Here is a plausible scenario that fits the detailed observations. The unlayered deposit was laid down at a time when net deposition exceeded net removal of polar material. Whether it was originally layered is unknown. Then, net removal exceeded net deposition and there was some net loss of material. Later, the situation again reversed and a set of layered deposits was laid down. There was another reversal and these layered deposits were partially removed. There was yet another reversal yielding a *second* set of layered deposits. The evidence for two separate sets of layered deposits is a topographical mismatch between the two sets called an *unconformity*, an example of which is shown in Figure 3.18. It has been suggested that the terracing need not be evidence for layering, but of a peculiar mode of erosion of an unlayered material. However, it is very difficult to account for the unconformity by this means. It seems more likely that the layers exist and reflect two series of atmospheric changes. Still more recently there has been another reversal to net removal, thus cutting the "valleys" and the pits. Whether there has comparatively recently been a switch to net deposition is unknown.

The fairly sudden reduction in weathering of the old craters occurred before all but the oldest surfaces were fashioned and probably occurred a good deal earlier than 3000 Ma ago. The likely associated decline in atmospheric mass is readily explained by loss to surface repositories and to a lesser extent to space. Cool temperatures on Mars would facilitate loss to such repositories.

Dendritic channels are confined to the southerly hemisphere, indicating that they may be relatively old. They can be placed in a time order by means of the small craters *in* them, and from evidence indicating whether they are older or younger than the terrain they cut. In the time-order they do not correlate clearly with the time-order position of volcanic events. The oldest dendritic channels are probably older than 3000 Ma. The younger ones may be only about 1000 Ma old or even younger making them younger than the terrain they cross. *If* they were carved in conditions in which liquid water was a stable phase, then the Martian atmosphere would need to contain about 200 times the present amount of CO_2 to enhance the greenhouse effect sufficiently to warm the surface. Such an amount could lie today in subsurface reservoirs. A *considerably* smaller amount of ammonia would do, because it is a very efficient producer of the greenhouse effect. However, there is no other evidence for ammonia ever having been present in the Martian atmosphere in appreciable quantities.

There are so few craters on the polar deposits that the intricate series of atmospheric changes that they record probably stretch back no more than a few hundred Ma. Some of the layers in the layered deposits may result from brief periods of volcanic activity, injecting ash into the atmosphere. But the main cause could be periodic variations in the orbital elements and axial inclination of Mars as in the Milankovitch theory of ice-ages on the Earth. The periods for Mars vary from about 0.095 Ma to about 1.2 Ma. The layering could correspond to the shorter periods and the changes from net removal to net deposition to the longer periods, though the detailed mechanism is unknown.

3.4.4 A Martian biosphere?

When it was established that the dark albedo features were not tracts of vegetation the case for the existence of a Martian biosphere lost its clearest-cut support. And yet were certain bacteria to be transferred from the Earth to Mars today it is likely that they would survive, particularly if they were shielded from the copious amount of solar uv radiation that reaches the Martian surface. It is therefore worth searching for a Martian biosphere. And the outcome of any search is significant regardless of whether life is found because it should help us understand how life originated on the Earth and which of the conditions on the Earth were essential for its development.

There are several types of observations of a planet that can be made to test whether it is supporting life.

You have seen that the Earth's atmosphere is largely a product of the biosphere both with respect to the major constituents, oxygen and nitrogen, and also with respect to certain minor constituents such as methane which would be even scarcer were it not produced at a high rate by the biosphere. The *far* smaller abundances of these gases in the Martian atmosphere indicates to some scientists that there is no Martian biosphere. But strictly speaking it only rules out a biosphere in which there are processes of tissue building and metabolism similar to those in the Earth's biosphere. It also remains possible that there *was* a Martian biosphere or that the present biosphere is small or dormant.

Orbiting spacecraft have yielded pictures of the Martian surface such that over most of the surface features larger than about 100 metres across are visible, and over a smaller fraction of the surface features larger than about 8 metres across are visible. Pictures obtained by the Viking Landers show detail near the Landers about the size of grains of sand and the size of large stones on the horizon a kilometre or so away. No organism living or dead nor any artefact has been seen. However, it is possible that such things have been seen and not recognized. For example, could any of those stones in Figure 3.16 be silicate "hats" protecting the organism beneath from solar uv radiation, the organism's tentacles extending downwards for several metres draining water from a dust/ice mixture? Probably not. But . . .

The Landers bore no microscopes to scrutinize the samples of bright-red dust that were drawn into the Landers. Chemical analyses of such samples were performed but revealed no compounds that typify organisms on the Earth. The most tell-tale group are the carbon compounds (section 2.1.1). The extremely small upper limits that the Landers placed on the abundances of carbon compounds indicate that *if* there are any Martian organisms based on carbon compounds then they are not at all widespread. The scarcity of carbon compounds also indicates a scarcity of the molecular detritus of organisms, though photodissociation by solar uv radiation can be adduced to explain this. It is, of course, possible that Martian biology is not based on carbon compounds though nothing is known about such an unlikely alien possibility.

The Landers also carried out experiments on the dust samples designed to detect the processes of tissue building and metabolism. All the results can just about be explained by chemical processes that do *not* involve organisms. These experiments rested on the very reasonable assumption that any Martian biology would be based on carbon compounds but also on the more questionable assumptions that the processes of tissue building and metabolism would be like some of those on the Earth. Moreover, the experiments were performed in the Landers in conditions that on the whole were warmer and moister than those on the Martian surface: this might have suited Earth bacteria but killed Martian organisms. Therefore, these experiments do not rule out a Martian biosphere. Moreover, the Landers have only scrutinized a tiny part of Mars. And they have not explored possible habitats such as the *interiors* of rocks: on the Earth, in the Antarctic, bacteria have been found several millimetres *within* rocks.

However, on balance, it must be admitted that Mars probably possesses no biosphere today. Whether it possessed one in the past is a more open question because of the possibility of less harsh climatic conditions in the past. But in the future it seems fairly certain that there will be life on Mars — us: unless humanity goes into a technological decline it seems likely that the first people will set foot on Mars within the next few decades.

What would it be like to roam the Martian surface?

3.4.5 Some highlights of a Martian tour

It might be the Earth somewhere in a dry antarctic valley. The Sun is just above the horizon and before us stretches a light-red dusty terrain strewn with boulders. A few hours ago this whole area was veneered with a glistening hoar frost which formed in the night when the meagre amount of water vapour in the atmosphere condensed on the ground. Then, after dawn the Sun warmed the ground, the hoar frost evaporated, at first to form a thin mist of ice crystals, and then it dissolved as the Sun rose higher.

This is 75°N, in the north polar region of Mars now emerging in late spring from the long Martian winter, the seasons on Mars being about double their length on the Earth. Last autumn, beneath the subdued and eerie whiteness of the polar hood the polar cap rapidly advanced. Now, it has as rapidly retreated. Only a week ago this valley still bore a few millimetres of CO_2 whitefall.

The Sun moves through the sky much as it does on the Earth in this nearly Earth-length day, and the noon altitude of the Sun varies with the seasons in much the same way as it does on the Earth, the two planets' axial inclinations being so similar. Even the size of the Sun is not strikingly different, about two-thirds of the angular diameter seen from the Earth, and the light it gives does not seem much dimmer to our eyes. The sky is more often yellow or orange than it is on the Earth and is otherwise a far deeper blue. The surface gravity is clearly less: here you weigh only 38% as much as on the Earth.

But in spite of the tempting similarity of the terrain don't be tempted to cast off your space suit. Out there lies, not antarctic discomfort, but rapid death. Your blood and other body fluids would boil at the low atmospheric pressure, and subsequently any body fluids that had not evaporated would be frozen solid. The highest temperature expected today is 163 K. Let's go somewhere else.

The view from the top of Olympus Mons is spectacular. In front of us the vast caulderas that pit the summit drop almost sheer for hundreds of metres. Behind us the dark-red corrugated flanks descend at a gradient of about 1 in 12 for hundreds of kilometres. The planes beyond are just visible on the horizon and are pitted here and there with impact craters.

It's a long hard 150 km walk to the far lip of the caulderas. Here the slopes stretch down as before until they meet a rolling white ocean of clouds that stretch away to the horizon. This today is the windward side of the mountain. The winds drive the atmosphere up the mountain's flanks into higher colder regions. The colder the atmosphere the less water it can hold in the vapour phase, and thus the excess water forms a rolling layer of small water ice crystals.

Overhead the sky is a deep purple, almost black. There is too little dust above us to give it a yellow or orange tint, and too little atmosphere to make it a brighter blue. The brightest stars above are visible all day, and when the Sun sets the stars shine with a brilliance and clarity unknown from the surface of the Earth, and stars too faint to be seen from the Earth are visible. Phobos and Deimos catch the eye with their strange motions against the stellar background. Phobos has a visible disc but Deimos is just a bright point of light.

Except for these satellites the night sky looks much the same as it does from the Earth. But among the extra stars are two that are so bright that they would be easily visible from the Earth and yet they are not. One is a brilliant blue-tinted point of light that outshines everything except Phobos and Deimos. The second is its companion, and is yellow and dimmer. Of course, the blue star is the Earth itself and its companion is the Moon. How sobering that all our lives, history, science and art are reduced to a point of light among so many others. How discomforting that home is a dot of light in a black sky, seemingly as unreachable as the stars.

3.5 Summary

Mars has for centuries been a source of fascination. Reality has fallen rather short of hope in relation to Martian biology, but Mars has not been disappointing in other respects.

The planet as a whole is probably made largely of the same sort of materials that make up the Earth, but in different proportions. In particular there is a lower abundance of metallic iron. Mars is differentiated, probably into a core and a mantle. It may be further differentiated, possibly including a surface crust. The interior is probably sufficiently cool for most or all of it to be solid today.

Mars has been less geologically active than the Earth, though it has been far from moribund. It also possesses an atmosphere, but it is a good deal less massive than that of the Earth and this has probably been the case for most of its history, though it is not certain whether Mars as a whole is depleted in volatiles in comparison with the Earth. The lower level of geological activity and the more tenuous atmosphere explain why numerous impact craters have survived on the Martian surface.

There is probably no Martian biosphere today, though there may have been a more clement time in the distant past when such a biosphere existed.

In the likelihood that Mars has had internal energy sources that, per unit mass of the planet, have been no greater than those in the Earth, then the lower level of geological activity would be the result of the smaller size of

Mars: the greater the surface area per unit mass the more rapidly a planet loses its internal heat and therefore the more rapidly it cools. The lower mass of the atmosphere is due, at least in part, to the lower surface temperatures.

3.6 Questions

1. List *six* major things that were unknown about Mars before 1877, and *six* major misconceptions that existed in 1965.
2. A coin 15 mm diameter is viewed from a range of 0.5 m. What is its angular diameter? *Roughly* how far away would it have to be placed to have the same angular diameter as Mars during the 1877 opposition?
3. List the five categories of information carried by electromagnetic waves.
4. What is the partial pressure of a gas which contributes 10^{-4} bar to the atmospheric pressure at the surface of a planet?
5. Outline how J_2 can be measured by placing a satellite in orbit around a planet.
6. Outline *two* ways in which the value of the polar moment of inertia (C) of a planet may be obtained.
7. What is meant by a planet being in hydrostatic equilibrium?
8. If Figure 3.9 represents a planet the size of Mars what is the altitude of the points P and Q on its surface?
9. Professor I. G. Norant thinks that Mars consists of 50% iron and 30% nickel beneath a crust of silicates and other materials, and that much of the metal is molten. Outline the evidence which indicates that he is wrong.
10. Two regions A and B on a planet's surface bear the same number density of medium and large impact craters but A bears fewer small craters and all of its craters are highly degraded. Another region C bears fewer craters than A of all sizes. Place the three regions in a time-order.
11. Summarize the evidence that Mars has been less geologically active than the Earth.
12. Over a series of oppositions a dark area on Mars is seen from the Earth to gradually get larger. Outline *what* is happening, and suggest a cause.
13. Give a plausible explanation of why there is no stratospheric "bulge" of temperature in the Martian atmosphere.
14. If you possessed a technology which could modify a planet, how would you give Mars an adiabatic lapse rate in the troposphere and a warm thermosphere?
15. Outline the evidence for accessible surface repositories of volatiles on Mars and for the loss to space of substantial quantities of volatiles in the past.

16. Summarize the evidence that there have been atmospheric/climatic changes on Mars.
17. In what ways would you try to establish whether there is any life on Mars?

4

VENUS

Venus, apart from the Sun and the Moon, is the brightest object in the sky and can cast a shadow perceptible to human vision. Because its orbit lies within that of the Earth it passes through inferior conjunctions rather than oppositions (Figure 1.6). It appears brightest when it is between inferior conjunction and a maximum elongation with about 30% of the Earth-facing hemisphere illuminated by the Sun. Venus is named after the Roman goddess of love. However, the adjective "Venusian" is not normally used, but instead "Cytherean" (Si-the-re-an) after Cythera, a Greek goddess of love.

4.1 The exploration of Venus

The first known observations of Venus through a telescope were made by Galileo in 1610. He noticed that Venus displayed phases like the Moon, that is, Venus varied from a thin crescent to full and back to a crescent again. Moreover, the variations of phase as Venus moved around its orbit were *not* in accord with the *Ptolemaic model* of the Solar System which originated with the Greek astronomer Ptolemy (*ca.* 100–*ca.* 170) a refined version of which was widely accepted in Galileo's time.

Figure 4.1 (a) shows the Ptolemaic model of the orbits of Venus, the Earth

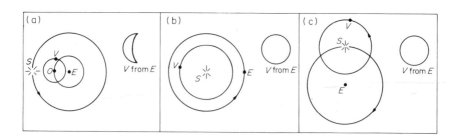

FIGURE 4.1
(a) Ptolemaic model of the Solar System. (b) The actual Solar System. (c) Tychonic model of the Solar System.

and the Sun. Several details have been omitted and I shall say nothing about the way Venus and the Sun move around their orbits, except that the point O on Venus's orbit always lies between the Earth (E) and the Sun (S), that V moves around O, and O and S move around the Earth. Maximum elongations occur when the angle SEV is a maximum, about 47°, in accord with observations.

The maximum phase of Venus in the Ptolemaic model is when the angle SVE is a minimum, and this occurs at the position shown in Figure 4.1 (a). The corresponding phase is also shown: Venus is noticeably under half full. Compare this with the true layout of the Solar System in Figure 4.1 (b) . Here the phase of Venus can approach full as shown and as observed by Galileo. The Ptolemaic model is therefore wrong. However, the true layout is not the only alternative. Figure 4.1 (c) shows Tycho Brahe's model, and you can see that this also accounts for the observed phases of Venus. It was other evidence which ruled out this model.

The earliest records of markings on the disc of Venus are by the Italian lawyer Franciscus Fontana (1585–1656) in 1643. He drew some dusky shadings. In the centuries that followed such shadings were shown to be transitory, and from these and other observations it became clear that Venus is completely and continuously covered in clouds. Figure 4.2 is a typical Earth-based view. The clouds are almost white in appearance, with the merest tinge of yellow, and their existence shows that Venus has an atmosphere to support them. The radius of Venus to the cloud tops is about 6120 km, which is only a bit less than the radius of the solid Earth. In the 1930s photographs at ultraviolet wavelengths displayed the visible shadings more clearly, and motions could be seen.

About 78% of the solar radiation incident upon Venus is scattered to space by the clouds. The remainder is absorbed within and beneath the clouds. Therefore, the scattered solar radiation carries no information about conditions beneath the clouds. Nor does it carry much information about the clouds to Earth because the clouds fail to impress on this radiation many meaningful signatures. The composition of the clouds was not known until spacecraft visited Venus. However, the scattered radiation does carry to Earth signatures impressed on it by the atmosphere *above* the clouds, and by the 1960s it had been established that this part of the atmosphere alone contained about 1000 times as much CO_2 as the whole atmosphere of the Earth. A trace of CO had also been detected and a trace of water with a column mass of about $0.01 \ kg/m^2$.

Such a small amount of gaseous water would be in equilibrium with clouds of condensed water ice crystals only if the cloud tops were extremely cold. In 1955 radiometry at planetary wavelengths established that the cloud-top temperatures were about 238 K, and it can be shown that if the clouds consisted of water ice crystals then there would be far more water in gaseous

FIGURE 4.2
Venus imaged from the Earth. North is at the top.

form in the atmosphere above the clouds than was observed. Clearly therefore the clouds do not consist of water, at least not in their upper reaches.

The first indication of conditions beneath the clouds came in 1956. In that year em waves from Venus at radio wavelengths greater than about 10 mm were detected by means of radio telescopes. Radiometric and spectrometric investigations of such waves indicated that the Cytherean source region was at a temperature of about 670 K. Moreover, radio waves at such wavelengths can readily traverse clouds and atmospheric gases and it was readily established that the Cytherean source region lay well below the cloud tops and was very probably the surface of Venus. Thus, the lower atmosphere and surface of Venus are *far* hotter than comparable regions of the Earth and Mars. Moreover if, as seemed likely, CO_2 was the dominant atmospheric constituent then the greenhouse effect of CO_2 could only maintain such temperatures if the column mass of atmospheric CO_2 was about 10^6 kg/m^2.

Up to this time many astronomers had regarded Venus as the Earth's twin, and that beneath the clouds, which were widely regarded as consisting of water, there were oceans and perhaps vegetation and animals. This idea was so firmly rooted that it was perhaps for this reason that these results from radio astronomy seem not to have been widely accepted at first. Thus, even though the cloud tops were known *not* to consist of water it was possible to believe that lower down they were of water and that deeper still lay a warm ocean-covered surface. However, through the 1960s the results of radio astronomy took root and oceanic Venus died. The high temperatures also killed off other views, such as that Venus was covered in oil.

By the 1960s the mass of Venus had long been known from its gravitational effect on the orbit of the Earth and of Mercury. This gives the ratio of Venus's mass to that of the Sun, and the mass of the Sun is obtained from the dominant elliptical part of the orbit of any planet just as the mass of a planet can be determined from the orbit of a satellite around it. Venus has no satellite and so this less direct method had to be used. The mass came to 0.82 M_E. Therefore *if* Venus has the same *mean density* as the Earth then its surface would lie about 400 km below the cloud tops.

The depth of the surface was established in the 1960s by *radar*. Radar studies of Venus began in 1961 and have revealed much of what we know about the Cytherean surface. Radar has also been a useful probe of other planets.

● *4.1.1 Radar studies of planets*

Radar is a technique whereby an object is investigated by beaming at it a short pulse of radio waves some of which are scattered by the object and picked up by a detector. In astronomy the transmitter is usually a radio telescope, and it usually acts also as the detector. The received waves are called *radar echoes* and they can carry the various categories of information that can be borne by any em wave, as outlined in section 3.1.2, though planetary temperatures cannot normally be extracted by radar. However, radar has the advantage over the passive receipt of radio waves that the pulses sent can be controlled and their nature and time of sending are known. *Radio waves* are used rather than em waves of other wavelengths because over suitable wavelength ranges they readily penetrate planetary atmospheres and because natural emissions at such wavelengths tend to be weak thus enabling the echoes to be readily picked out from the natural background.

Figure 4.3 illustrates the types of information that can be obtained about a planet using radar which could not be, or not as readily be, determined by the passive receipt of naturally occurring radio waves.

Figure 4.3 (a) illustrates how radar can yield the *radius* of a planet. A radar pulse scattered back by the planet is spread over a distance range which exceeds that of the transmitted pulse by an amount equal to the radius of the planet. (Note that waves scattered in directions other than back towards the source have been omitted, and that the transmitted pulse would cover the whole planet.)

Figure 4.3 (b) illustrates that the axial spin rate, and therefore the axial period of a planet can be determined from the spread of *wavelengths* induced in the echo by the motion of the surface. This spread arises from the *Doppler-shifts* mentioned in section 3.1.3. The planet as a whole will be moving with respect to the Earth, and this will shift all wavelengths in the pulse. But the edge of the disc moving towards the Earth will result in a slightly different shift from that of the disc moving away from the Earth. Thus, as well as an overall shift in the wavelengths there will also be a spreading of wavelengths, and from the measured spread the difference in the relative speed of the extreme edges of the disc, P and Q in Figure 4.3 (b), can be deduced. Clearly

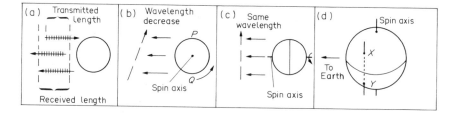

FIGURE 4.3
(a) Radar pulse broadening by a planet. (b) Doppler spreading of wavelength by a planet's rotation. (c) *No* Doppler spreading in this orientation. (d) Illustrating the range-Doppler method.

this gives the axial spin rate if the radius of the planet is known, and it is also necessary to know the axial orientation of the planet with respect to the Earth. The need to know the orientation is illustrated by comparing Figure 4.3 (b) and (c): in Figure 4.3 (c) the axis points at the Earth and there is no wavelength spread in the echo induced by the axial rotation. In Figure 4.3 (b) the spread is maximum. By making radar observations of a planet at several points in its orbit, and therefore at several axial orientations with respect to the Earth, it is, however, possible to obtain the axial orientation, and hence the axial period can be deduced. Moreover, from the axial orientation the axial inclination of the planet is readily determined. In the case of planets like Venus, where surface features are obscured, this is an important way of determining the axial period *and* axial inclination.

Radar can provide a more direct determination of axial period if there is a feature in the echo that can be inferred to have come from a surface feature on the planet, such as a mountain or a region with a particular composition or texture. By observing many separate echoes the movement of such features through the echo can be discerned, thus yielding the axial period.

Radar can also yield images, but whereas a radio telescope or a group of such telescopes can yield an image of an object in an analogous way to a conventional telescope, a radar pulse from the Earth cannot be made narrow enough to select anything other than the whole disc of the planet, except for the Moon where small areas can be examined because of its large angular diameter. Therefore a special technique for radar imaging has been developed, as illustrated in Figure 4.3 (d). This shows the simple case of a spin axis which is perpendicular to the line of sight. The line XY is parallel to the spin axis, and the extension of this line above and below the planet's surface allows for surface topography, the surface shown in Figure 4.3 (d) being the *mean* surface of the planet.

All contributions to the pulse echo from anywhere along the line XY, which in practice means the two patches of the planet's surface on this line, corresponds to unique values of a certain two quantities which no other points on the surface possess. The two quantities are the time taken for the waves to reach a patch and return to the Earth, which gives the *range* of the patches, and the Doppler-shift of such waves, which gives the motion of the patches. The echo comes from the whole disc, but because of this uniqueness the echo can be unscrambled to reveal the contributions to it from every pair of patches of planetary surface exemplified by X and Y.

Several sophisticated techniques are available, many of which involve more than one radio telescope receiving the echo, by means of which the ambiguity between the two patches can be removed, and the topography, surface texture and information about the composition of a planet's surface can be established.

Radar techniques based on this division of a surface by range and by Doppler-shift are called *range-Dopper methods*.

The need to use range-Doppler methods is obviated where radar equipment is carried near to a planet by means of a spacecraft. In this case the radar beam can cover a small area of surface at any one time. It is, however, essential to know where the spacecraft is. The *distance* from the Earth to the spacecraft can be determined very accurately from the time interval between sending a radio signal to the spacecraft, which it receives and immediately acknowledges by sending a signal to the Earth with its own radio, and the receipt on the Earth of the acknowledgement. The *speed* of the spacecraft along the line of sight from the Earth to the spacecraft can be obtained from the Doppler-shift of the known on-board transmitter wavelength. These observations of distance and speed can be coupled with general knowledge of the way a small probe behaves in the gravitational fields of bodies in the Solar System to determine the spacecraft's position with astonishing accuracy, typically to a few hundred metres at a range of hundreds of *millions* of *kilo*metres. Any occultation of the spacecraft by the planet will supplement this positional information.

4.1.2 Radar studies of Venus

Radar studies of Venus have established that the mean radius of the surface beneath the clouds is 6052 km, which places the surface about 70 km beneath the visible cloud tops. Radar studies also yielded a more accurate value of the mass of Venus by determining with great accuracy the distances between the Earth, Venus and Mercury, and hence establishing with great accuracy the variations in the orbital elements of these planets.

The sidereal axial period and the axial inclination were also established by radar, and the earliest values are close to the presently accepted best values (Table 1.3) of -243.01 days and $177.8°$ respectively. The minus sign attached to the period indicates that the axial rotation is *retrograde*, and this is also indicated by an inclination in excess of $90°$, though the angle between the spin axis and the perpendicular to the Cytherean orbital plane is only $2.2°$.

The retrograde rotation of Venus came as a considerable surprise. Previously, on the basis of very sparse data, estimates of the sidereal axial period ranged from $+15$ days to $+224.701$ days, the upper limit equaling the sidereal *orbital* period. This equality corresponds to *synchronous rotation*. This is illustrated in Figure 4.4 (a), and you can see that Venus would keep the

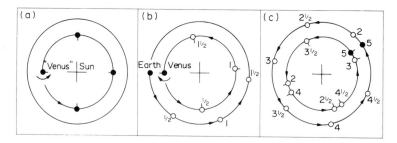

FIGURE 4.4

(a) Synchronous rotation, once thought to apply to Venus. (b) and (c) The actual rotation of Venus, the numbers indicating the number of axial solar periods of Venus lapsed. The line on Venus is fixed to its surface at the equator.

same face towards the Sun. Note that for synchronous rotation the orbital and axial periods not only have to be equal but also both prograde or both retrograde. Synchronous rotation results from so-called *tidal* interactions between bodies, to be outlined in section 6.1.1. The *actual* rotation of Venus is illustrated in Figure 4.4 (b) and (c). Between one inferior conjunction and the next Venus rotates *almost* exactly five times as viewed from the Sun and four times *as viewed from the Earth* and therefore Venus *almost* presents the same face to the Earth at each inferior conjunction. If the sidereal axial period of Venus were -243.16 days then, except for the slight effect of the eccentricity of the orbits of Venus and the Earth, Venus *would* present the same face to the Earth at each inferior conjunction. If such an exact situation existed it would be even harder to understand than the approximate situation which does exist. The slow, retrograde axial rotation of Venus is a puzzle to which I shall return in section 4.2.

Earth-based radar studies have determined the topography and the texture of about 25% of the Cytherean surface, but radar studies from spacecraft have been the main source of such information. Table 4.1 lists the successful spacecraft missions to Venus. I shall return to these surface studies in section 4.3.

TABLE 4.1
Successful spacecraft missions to Venus

Date of Cytherean encounter	Name of mission	Type of mission	Country of origin
14 December 1962	Mariner 2	Fly-by	USA
19 October 1967	Mariner 5	Fly-by	USA
15 December 1970	Venera 7	Lander	USSR
22 July 1972	Venera 8	Lander	USSR
5 February 1974	Mariner 10	Fly-by*	USA
22 October 1975	Venera 9	Orbiter + lander	USSR
25 October 1975	Venera 10	Orbiter + lander	USSR
4 December 1978	Pioneer Venus 1	Orbiter	USA
9 December 1978	Pioneer Venus 2	Four landers	USA
21 December 1978	Venera 12	Lander	USSR
25 December 1978	Venera 11	Lander	USSR
1 March 1982	Venera 13	Fly-by + lander	USSR
5 March 1982	Venera 14	Fly-by + lander	USSR

* Also flew by Mercury.

4.2 The interior of Venus

The best value of the mass of Venus has been obtained from the orbit of the Pioneer Venus Orbiter (Table 4.1), and is 0.8150 M_E. The mean density of Venus is therefore 5240 kg/m^3, rather less than that of the Earth (5520 kg/m^3) and substantially more than that of Mars (3940 kg/m^3). When allowance is made for the rather lower mass of Venus than of the Earth, and the consequent lower degree of internal compression, then the uncompressed densities of the Earth and Venus would be similar. Therefore, Venus and the Earth could have similar internal structures and compositions, with iron cores and silicate and oxide mantles.

It has not yet proved possible to constrain the variation of density with depth in Venus. From the orbit of Pioneer Venus Orbiter a value of J_2 (section 3.2.1) has been obtained, but it is about 200 times smaller than J_2 of the Earth, and consequently is known with poor accuracy, to about 40%. J_2

is in any case a weak constraint on density models and when it is this poorly known it is useless. The polar moment of inertia C is a far more powerful constraint. For Venus the only practical way of obtaining C is by calculation from any two of J_2, T_a (the axial period) and f (the flattening) making the reasonable assumption that Venus is in hydrostatic equilibrium. J_2 is poorly known, and only an upper limit exists for f such that the polar radius is within 0.24 km of the equatorial radius. The Earth is over 100 times more flattened than this upper limit. Therefore, though T_a is known, J_2 and f are too poorly known for the calculated range of possible values of C to provide a useful constraint on density models.

The small value of J_2 and the small upper limit of f are consistent with hydrostatic equilibrium because of the slow spin of Venus and the absence of any large satellite to increase J_2 and f.

Weak magnetic fields have been detected associated with the ionosphere and perhaps with surface rocks, but not any magnetic field of internal dipole origin. The upper limit on the internal magnetic dipole moment is only 0.00005 of that of the Earth. The slow spin of Venus had led many scientists to expect that the dipole moment would be small (section 2.1.7) but perhaps not as small as this. In section 4.3 you will see that there is some evidence that Venus is not as geologically active as the Earth, which could indicate a cool interior right down to any iron core, which could largely be solid and thus explain the small dipole moment. However, given the similar masses and densities of Venus and the Earth it is hard to understand how their interior temperatures could be so different. An alternative explanation is that the solid core at the centre of the Earth's liquid iron core is an essential source of energy for a powerful magnetic dipole moment (section 2.1.7). Venus could lack such an inner core because of the lower central pressures corresponding to its lower mass. It is also possible that Venus is not divided into a core and a mantle. However, the density of Venus could then only be matched if its composition, in terms of the abundances of elements, were markedly different from the Earth.

One way of explaining the slow spin of Venus is through the tidal interaction (section 6.1.1) between Venus and the Sun. Such an interaction acts through any departures from a spherically symmetrical distribution of mass in Venus. Such departures, as you have seen in relation to J_2, are small and therefore this tidal interaction is weak. It is far weaker between Venus and the Earth and therefore the nearness with which Venus keeps the same face to the Earth at superior conjunction (section 4.1.2) is probably a coincidence. The weakness of the tidal interaction with the Sun may mean that even after 4600 Ma synchronous rotation is yet to be achieved, and that in the distant future the retrograde rotation will cease and then the requisite slow prograde rotation will build up.

But this does not explain why Venus is spinning *retrograde*. However, the present winds on Venus are acting to *increase* the retrograde rate of spin, and thus the present rotation of Venus may be a balance between the tidal forces, which depend on the Sun's gravitational field, and the winds, which depend on solar radiation.

It is also possible that the rotation of Venus was reversed by the impact of a large body during or shortly after its formation: more on this in Chapter 15. In Chapter 15 you will also see that some theories of the origin of the Solar system carry the possibility that the accretion of the planets could have led to one or two with retrograde spins.

4.3 The surface of Venus

The topography of about 93% of the Cytherean surface has been mapped by the Pioneer Venus Orbiter. The principal technique was to transmit radar pulses vertically downwards and measure the time lapsed before the echo was received. This gave the height of the spacecraft above the surface. The corresponding distance of the spacecraft from the centre of mass of Venus was determined from the spacecraft's orbit. By subtracting from this distance the spacecraft's height the distance of the point on the surface from the centre of mass is obtained. The zero of altitude was chosen as a *sphere* of radius 6051.0 km, with centre at the centre of mass. A sphere is used because the surface of Venus corresponding to hydrostatic equilbrium is very nearly spherical (section 3.3.1).

The radar also yielded the scattering properties of the Cytherean surface. Features about as small as 30 km across could be detected, roughly equivalent to viewing the Moon from the Earth with an angular magnification of about 4, readily obtainable with binoculars. Altitude differences of about 200 metres could be detected.

These data were supplemented by the radar "looking" sideways at the surface, a rather subtle technique which I shall not describe.

The outcome is shown in Figure 4.5. The surface is solid, and the various features are named, in the main, after women in mythology and in science. A prominent male interloper is James Clerk Maxwell (1831–1879) the British physicist after whom has been named Maxwell Montes the tallest mountains on Venus, reaching an altitude of nearly +12 km. Figure 4.5 is broadly consistent with data from Earth-based radar some of which has yielded data on a smaller scale in equatorial regions than that from the Orbiter.

Three types of region have been recognized on the Cytherean surface.

First, there are *rolling planes*, which comprise about 65% of the imaged surface area and which lie approximately between altitudes 0 km and +2 km. They display linear troughs, sets of parallel ridges and troughs, and numerous roughly circular features from about the lower size limit of 30 km across up to about 1700 km across. Their interiors are "radar-dark" in that

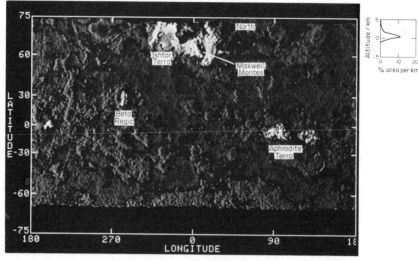

FIGURE 4.5
Venus if it were devoid of atmosphere, and the corresponding hypsometric distribution.

they scatter radar more weakly than the surface outside them, and some of the larger ones have central mounds. It is not known whether they are of volcanic or impact origin, or a mixture of the two, though the relative numbers of different sizes are similar to the impact craters in the more heavily cratered areas on Mars and the Moon.

Second, there are *highlands*, which comprise about 8% of the imaged surface area and which lie approximately between altitudes +2 km and +12 km. There are three main highland regions (Figure 4.5): Ishtar Terra, which is a plateaux; Aphrodite Terra, which is rougher, and bears wide linear depression a few kilometres deep; and Beta Regio, which could possibly be two shield volcanoes.

Finally, there are the *lowlands*, which comprise about 27% of the imaged surface area and which lie approximately between altitudes −2 km and 0 km. They consist of roughly circular depressions and of more elongated depressions, and are radar-dark. Within these depressions very few topographic features have been seen.

No global system of oceanic ridges, subduction troughs and mountain ranges at continental borders has been seen at the 30-km resolution achieved by the Orbiter. Thus, a global system of Plate Tectonics is *not* revealed by the topography, though it would require resolution of about 15 km to firmly rule it out. Such a resolution would also establish whether impact craters and volcanoes exist.

A clearer indication that there is no global system of Plate Tectonics is the hypsometric distribution of Venus, also shown in Figure 4.5. This is evidence *against* there being two types of crust on Venus.

On the other hand, the Orbiter has established that most, perhaps all, of the Cytherean surface is close to isostatic equilibrium, which indicates that there may be a fairly thin lithosphere plus an asthenosphere, and this raises the possibility of considerable geological activity. Also, the centre of figure of Venus is offset from the centre of mass by only about 400 metres, a far more Earth-like than Mars-like amount, and which is consistent with an interior sufficiently warm to eliminate any lopsided density distributions.

Several spacecraft have landed on the Cytherean surface (Table 4.1) and some of the USSR Venera spacecraft have carried cameras. Figure 4.6 shows two pictures released. One is from Venera 9 and the other from Venera 10. Both of these spacecraft landed near the eastern edge of Beta Regio (Figure 4.5). You can see that the surfaces are littered with boulders, a degree of roughness consistent with radar scattering data. Boulder-strewn dusty surfaces have also been revealed by Veneras 13 and 14, which landed on rolling planes a few thousand km to the south-east of Beta Regio. They also returned the first *colour* pictures from the surface, which show that the rocks are tinted brown. The angularity of the rocks indicates chemical weathering by the atmosphere.

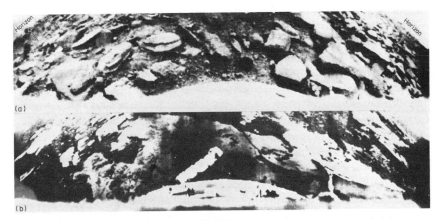

FIGURE 4.6
The surface of Venus (a) imaged by Venera 9, (b) imaged by Venera 10. The curvature of the surface is an artefact of the wide-angle view.

Analyses of the rocks at the Venera 13 and 14 sites are consistent with a basaltic composition. Veneras 9 and 10 carried out less revealing analyses, and so too did Venera 8, which landed rather further from Beta Regio than the other Veneras. Nevertheless, the analyses from Veneras 9 and 10 are also consistent with basaltic composition whereas the analyses from Venera 8 are less consistent, and may indicate a substantially different kind of rock. It is likely that the rocks of basaltic composition have a volcanic origin, and could have flowed as basaltic lavas from any Beta Regio volcanoes. In this case the rocks at the Venera 8 site may be non-basaltic because of the greater distance of this site from Beta Regio.

But though Venus does not seem to be geologically moribund it seems likely that it has not been as geologically active as the Earth and that no global system of Plate Tectonics has ever developed. If indeed this is the case then the question arises of why the Earth and Venus, which have similar masses and densities, differ so much in this respect. A comparatively cool Cytherean interior could account for the difference, though this raises the further question of how the Cytherean interior could be so much cooler than that of the Earth: has the tidal heating of the Earth's interior by the Moon been important (section 2.1.16), or is the Cytherean interior depleted in the isotopes which can supply copious amounts of radioactive heat, namely, uranium, thorium and potassium? Or is there some other explanation?

Clearly, the question of whether Plate Tectonics has developed on Venus is of considerable interest.

4.4 The atmosphere and surface volatiles

4.4.1 The atmosphere today

The composition of the Cytherean atmosphere in terms of relative numbers of molecules is as follows: 96% CO_2; 3.5% N_2; 0.015% SO_2; about 0.01% H_2O in the lower atmosphere; 0.007% Ar; plus traces of several other gases. Oxygen (O_2) has not been detected in the lower atmosphere, though small quantities have been detected in the upper atmosphere and result mainly from the photodissociation of CO_2 by solar uv radiation. This also produces most of the ions in the Cytherean ionosphere. This extends upwards from an altitude of about 160 km. Up to about 200 km the dominant ion is O_2^+ and at higher altitudes it is O^+.

The vertical structure of the atmosphere, averaged over the surface and over time, is shown in Figure 4.7. The atmosphere is massive, 103×10^4 kg/m^2, which dwarfs the 1.04×10^4 kg/m^2 of the Earth's atmosphere. The average surface temperature on Venus is also much higher than on the Earth, 730 K instead of 288 K. The troposphere extends up to an altitude of about 60 km, and in its lower 50 km or so the lapse rate is adiabatic. There is no stratosphere "bulge" of temperature for much the same reason as there is no such bulge in the Martian atmosphere (question 3.13). And for much the same reason as for Mars the thermosphere of Venus only reaches modest temperatures (section 3.4.1). However, the thermospheres of Mars and Venus have comparable temperatures only for the Sun-facing side of Venus. At night the thermosphere of Venus cools far more than that of Mars, and reaches the low temperatures indicated in Figure 4.7. Only a small part of

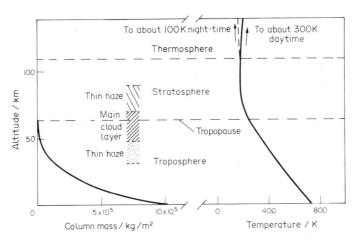

FIGURE 4.7
The vertical structure of the Cytherean atmosphere.

this difference between the two planets is due to the much longer night on Venus: the main reason is unknown. Moreover, the Cytherean thermospheric temperatures are not very sensitive to changes in solar uv radiation. Therefore, the Cytherean thermosphere is not very well understood. The night-time thermosphere is sometimes called the *cryosphere* (= "cold sphere").

Figure 4.7 also indicates typical altitude ranges of the clouds and hazes. These clouds are unbroken, and never, for example, look like the billowy cumulus clouds of the Earth. Moreover, the main cloud base lies at an altitude of about 48 km, which by Earth standards is a *very high* cloud base. Even the haze beneath the cloud lies above the high altitude of about 31 km beneath which the Cytherean atmosphere is clear most of the time.

Detailed studies, mainly by spacecraft, have revealed the existence of at least *three* types of cloud particle. First, there are tiny particles a few tenths of a *micron* in diameter (1 micron $= 10^{-6}$ metres). The observations indicate that they probably consist of nearly pure *sulphuric acid* (H_2SO_4). These are present in the main cloud and in the hazes above and below, but though they are the dominant constituent of the hazes they are a minor constituent of the main cloud. Second, there are larger particles about 2 microns diameter largely confined to the main cloud each particle probably consisting of a liquid droplet of rather impure H_2SO_4 with up to 20% of other substances. The third type seem to be solid, and are largely confined to the lower half of the main cloud. They are about 8 microns in diameter and probably consist largely of chlorides, that is, of compounds containing chlorine. On the Earth such compounds are largely confined to the oceans, common salt (sodium chloride) being an abundant example. This third type is the dominant constituent of the main cloud and also account for most of the opacity. However, the particles in even the most opaque Cytherean clouds account for only a tiny fraction of the atmospheric volume occupied by the cloud, and therefore the cloud particles, as on the Earth, are minor atmospheric components.

The sulphuric acid is probably the result largely of chemical reactions between H_2O and SO_2 high in the atmosphere above the haze and clouds, reactions aided by solar uv radiation and by the low temperatures. Tiny droplets form and constitute a thin rain which feeds the hazes and clouds. This thin rain continues beneath the clouds, where most of the droplets evaporate. In this subcloud region there is little solar uv radiation and the temperatures are fairly high, and therefore the molecules of H_2SO_4 revert back to H_2O and SO_2. Upward convection in the troposphere carries H_2O and SO_2 into the atmosphere above the clouds thus completing the cycle. In passing through the hazes and clouds some of the gaseous H_2O would dissolve in the H_2SO_4 particles, thus diminishing the H_2O abundance in the atmosphere above the clouds. But though there is a greater abundance of H_2O in the atmosphere *beneath* the clouds it is still a very minor constituent, and the quantity is poorly known.

Plausible schemes also exist for producing the other constituents of the cloud particles.

Fine drizzles and thin smokes of cloud particles can reach the Cytherean surface. Moreover, thunderstorms are known to occur in the clouds and these could generate H_2SO_4 droplets large enough to constitute a heavy shower, though the torrid surface would soon evaporate any puddles of acid.

The high surface temperatures of Venus are a consequence of the greenhouse effect. If the Cytherean greenhouse effect were "switched off" but everything else were left the same the average surface temperature would fall from 730 K to 230 K. For the Earth, the corresponding temperatures are 288 K and 253 K. Thus, though Venus is closer to the Sun than the Earth its surface temperature would actually fall *below* that of the Earth. This is because of the higher Bond albedo of Venus. Indeed, only about 4% of the solar radiation available at the top of the Cytherean atmosphere reaches the surface. Nevertheless, the greenhouse effect is so efficient that this energy becomes trapped to the extent that it drives up the surface temperatures to the observed values. Recall from section 2.2.3 that this trapping results from the atmosphere being more opaque to radiation at planetary wavelengths than at solar wavelengths. Such efficient trapping *cannot* be produced by the CO_2 alone, in spite of the enormous mass of CO_2 in the atmosphere. This is because CO_2 is fairly transparent over certain wavelength ranges at planetary wavelengths. Radiation could escape through these "windows" in sufficient quantities to greatly reduce the greenhouse effect below that which exists. It is the blocking of these windows by SO_2, by H_2O and by the clouds that greatly increases the greenhouse effect.

When viewed at visible wavelengths the Cytherean clouds are rather featureless, and are tinged yellow, perhaps by small quantities of sulphur. At uv wavelengths considerably more detail is seen, as illustrated in Figure 4.8. The dark markings correspond to regions in the clouds which scatter less solar uv radiation than elsewhere. The substances responsible for this are not known, though they are clearly unequally distributed and this is somehow linked to the circulation of the Cytherean atmosphere though the mechanism is unclear.

The dark markings in Figure 4.8 are subject to changes in detail but persist in broad outline, indicating fairly stable patterns of circulation. Detailed scrutiny of these and other markings reveals that, at the cloud tops, the atmosphere moves around Venus predominantly in an east–west (zonal) direction and in the same direction that Venus rotates, taking about 4 to 5 days to encircle Venus once. At the equator the corresponding zonal wind speed (at 60 km altitude) is about 100 m/s. There is also an equator to pole flow at cloud-top level, but at only 5–10 m/s.

FIGURE 4.8
The clouds of Venus, imaged at ultraviolet wavelengths by Mariner 10 from a range of 720 000 km. North is at the top.

The winds at lower altitudes have been measured at a few locations by landers as they descended through the atmosphere. Over most altitude ranges the wind speeds decline with decreasing altitude, and at the surface are only about 1 m/s. The increase of speed with altitude could be the result of the upward transfer, by convection in the troposphere, of energy of motion: as such energy is transferred from the high-density regions at low altitudes to the low-density regions high up then the low-density material has to move faster to carry this energy.

The north–south winds vary in direction with altitude, though the zonal winds are in the direction of rotation at all altitudes.

The circulation pattern can be thought of as made up of several contributions. One of these is called the *thermal tide* and consists of a flow from the region beneath the Sun to the opposite side of the planet. By this means the Sun-facing regions become cooler than they would otherwise be. Thermal tides occur in the atmospheres of other planets. In the case of Mars they make a very *large* contribution to the overall circulation of the atmosphere whereas in the case of the Earth the contribution is small. In the case of Venus thermal tides make a very large contribution and it can be shown that it is this component of the overall circulation that acts to speed up the axial spin of Venus (section 4.2).

The thermal tide may also be responsible for dim ultraviolet radiation observed to come from the middle of the night hemisphere. The thermal tide could carry molecules generated by solar uv radiation to the night side where they revert to other forms with the emission of uv radiation. A similar mechanism may also yield the *Ashen Light*, a faint radiation at visible wavelengths and seen from the Earth to emanate from the night hemisphere.

The circulation of the massive atmosphere of Venus results in a very small difference in surface temperatures between the equator and the poles and between the night and day sides, a few Kelvin in both cases. Moreover, the axial inclination and orbital eccentricity of Venus are so small that seasonal variations are negligible.

4.4.2 Surface volatiles today

If mixtures of certain types of rocks including carbonates are held in a sealed evacuated chamber at a temperature equal to the surface temperature of Venus then CO_2 is given off and the pressure builds up until an equilibrium is reached in which the pressure of the CO_2 in the chamber equals the pressure of CO_2 at the surface of Venus. This suggests that the atmospheric CO_2 on Venus is in equilibrium with these types of surface rock and that there exist on Venus accessible surface repositories of CO_2 mainly in the form of carbonates. At the much lower surface temperatures on the Earth most of the CO_2 remains in the carbonates and therefore the partial pressure of atmospheric CO_2 is low. Nevertheless, on Venus it remains possible that no such surface repositories exist and that the correspondence of pressures is coincidental.

Sulphur compounds are probably present at the Cytherean surface, and the partial pressure of SO_2 and other gaseous sulphur compounds is probably determined by equilibrium with these surface repositories, as suggested for CO_2. Likewise, oxygen-rich compounds at the surface probably control the oxygen abundance in the lower atmosphere, though such oxygen has yet to be detected. Surface deposits may control the atmospheric abundances of other trace gases too.

It is not known how large these surface repositories are. But in the case of nitrogen, which can exist as nitrates and nitrites, the surface temperatures are so high that the amount retained in such repositories must be small compared to the amount in the atmosphere.

The high surface temperatures also mean that there can be no accessible repositories of water in which the water exists as H_2O whether as a separate phase or in some molecular combination. Water could, however, be present in the hydroxyl form (section 2.1.1). For example, on the Earth water in hydroxyl form occurs in clays, and if similar clays were exposed on the surface of Venus then the abundance of water in the atmosphere would be roughly that observed. However, the abundance of atmospheric water is also consistent with a gas-phase equilibrium with the sulphuric acid and water in the cloud droplets and it is therefore far from certain that significant accessible repositories of water exist, even with water in the hydroxyl form. In any case, it is unlikely that any such repositories contain anywhere near the amount of water in the Earth's oceans. This likely scarcity of water on Venus is one of the main topics of the next section.

4.4.3 The evolution of the atmosphere

If the Earth were to be moved today into the orbit of Venus then the ice caps would melt, the oceans would start to evaporate, the carbonate rocks would begin to yield up CO_2, and various other surface reservoirs would begin to lose volatiles. The greenhouse effect would be enhanced, and this would further raise the surface temperatures and thus increase the amounts of volatiles lost by the surface repositories. Before very long on a planetary time scale the surface temperatures would reach Cytherean values. The increasing role of the greenhouse effect in this transition has been called the *runaway greenhouse effect*.

But in one important respect the Earth and Venus would differ: the atmosphere of the new Earth would contain at least as much water as there is CO_2 in the present Cytherean atmosphere. Was Venus born dry, or did it become dry?

Water can be lost through dissociation by solar uv radiation and the subsequent loss of hydrogen to space from the exosphere. Without this loss an equilibrium would be reached in which only a small proportion of the water in the atmosphere would be dissociated. However, the temperature today at the base of the Cytherean exosphere rarely exceeds about 300 K, which is too low for the thermal escape of very much hydrogen even over 4600 Ma, and chemical escape might not make a lot of difference. But atmospheric conditions may have been different in the past. An indicator is the abundance of the rare isotope of hydrogen, deuterium D, compared to the common isotope ^{1}H. D is twice the mass of ^{1}H and therefore ^{1}H will normally escape faster, enhancing $D/^1H$. The $D/^1H$ ratio in the Cytherean atmosphere today is about 100 times that on the Earth, indicating more ^{1}H on Venus in the past, though still a good deal less than on the Earth. However, it can be shown that the runaway greenhouse effect, had it acted on an Earth-like quantity of water, can perhaps lead initially to a copious escape of hydrogen with *no* enhancement of $D/^1H$ during this process. It is thus possible that Venus *did* once possess a large quantity of water. Conversely, evidence that no more than a modest quantity of water has been lost comes from considering the fate of oxygen liberated. It would not escape to space in appreciable quantities, and so combination with surface materials provides the only repository. However, it is difficult to believe that sufficient quantities of suitable substances could have appeared at the Cytherean surface to mop up anywhere near the amount of oxygen that would be liberated from an Earth-like quantity of water. Indeed, chemical evidence from the sulphur-bearing gases in the lower atmosphere of Venus can be shown to indicate that the surface rocks today do *not* contain the maximum amount of oxygen possible.

However, if Venus has always had as little water as it seems to have today then it may never have been able to achieve such high surface temperatures. The luminosity of the Sun in the early history of the Solar System was probably only about 70% of the present value. Under these conditions most of the CO_2 presently in the Cytherean atmosphere would have resided in carbonate rocks. As the solar luminosity rose some of this CO_2 would have been liberated, but it can be shown to be unlikely that sufficient CO_2 would be liberated to yield present conditions. But a small *extra* quantity of water could enhance the greenhouse effect to the point where present conditions would result. This extra water could subsequently be lost without reversion to cooler conditions, and such a relatively small quantity could be lost by uv dissociation as outlined above. Thus, the present water is sufficient to help *sustain* the temperatures but not to *produce* them. A trace of ammonia could also have sufficiently enhanced the greenhouse effect, though there is no other evidence that such a trace ever existed.

It is possible that, perhaps for as much as its first 2000 Ma, Venus enjoyed clement conditions, with small open bodies of water, and rivers. This water could have left features visible today, such as dendritic channels. Unfortunately, the spatial resolution of the best radar images so far obtained is too poor to have revealed such tell-tale signs. It is also possible that life developed, and if the rate of evolution was about twice as rapid as on the Earth then intelligent beings may have been among the organisms roasted to death as the Sun's luminosity rose, the small oceans evaporated, and the CO_2 emerged from the rocks.

4.5 A brief (and well-protected) visit to the Cytherean surface

One of the most familiar things is the surface gravity, about 0.91 of that of the Earth. Also, in spite of the canopy of thick cloud it is not dark during the day, but is rather like being beneath very thick thunder clouds on the Earth, though the cloud base on Venus is far higher, and, as revealed by Venera 14, the sky is orange. At night the landscape glows a dull red, which emanates from the surface of Venus: the rocks, hotter than a domestic oven, glow just enough to be seen at night. Lead melts.

The thick, unbreathable atmosphere carries traces of what to us are corrosive and poisonous substances. The surface winds are fairly light, but from time to time can raise dust. Now and then a fine drizzle of sulphuric acid droplets may fall, plus other cloud particles, and the occasional thunderstorm high overhead can give a heavier rain. But otherwise the weather on Venus varies very little, either with place or with time.

The landscape is a stony desert, smeared here and there with sulphur compounds. There are mountains, canyons, perhaps impact craters, and perhaps volcanoes, some of which may even be still active. And the ground may shake to the occasional "Venusquake".

The horizon on this Earth-sized world lies at about the same distance as on the Earth. But *unlike* on the Earth we can see well beyond the horizon. The reason is illustrated in Figure 4.9 (a). It can be shown that the decrease of

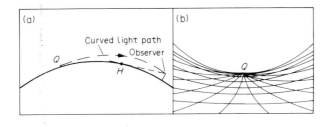

FIGURE 4.9
The view over the horizon on Venus.

density with height which occurs in all planetary atmospheres results in the bending of em waves including light. For this reason we can see a little way beyond the horizon H even on the Earth. But on Venus, because of its much more massive atmosphere the bending is more marked and we are able to see *well* beyond the horizon, particularly because the lower atmosphere of Venus is usually very clear. If one were in the middle of a flat plane, conveniently marked with a square grid, then the outcome would be as in Figure 4.9 (b). The bending of the light is sharper than the curvature of Venus, and the impression is therefore of standing at the bottom of a shallow bowl.

If there were any Venusians (perhaps made of asbestos!) they would have no immediate knowledge of other worlds. High above them the constant roof of clouds would merely lighten and darken in a long solar axial period of 116.8 days. They would have to await the invention of radio astronomy to "see" beyond the clouds.

4.6 Summary

Cloud-shrouded Venus remained a mystery until the advent of radar and visitation by spacecraft. It is still a relatively poorly explored planet.

Its size and mass are similar to the Earth, but too few data exist for it to be known whether its interior composition and structure are as similar. There is evidence for some geological activity on Venus, and though it is unlikely that Plate Tectonics ever got going this conclusion is not yet beyond doubt. *If* it has been considerably less geologically active than the Earth then the reason is not clear, though one or two plausible suggestions have been made. The reason for the low upper limit on the internal magnetic dipole moment is not properly understood.

The Cytherean atmosphere is far more massive than that of the Earth, and the surface temperatures are considerably higher. This is because of Venus's smaller distance from the Sun which has resulted in a runaway greenhouse effect. However, the present scarcity of water in the atmosphere is a puzzle. It seems likely that there was a greater abundance of water in the past, though it is unlikely ever to have been as large as the abundance of water on the Earth. I shall return to this puzzle in the next chapter.

4.7 Questions

1. Outline how observations of Venus ruled out the Ptolemaic model of the Solar System.
2. State briefly, in relation to planetary observations, the advantages of radar over the receipt of naturally emitted radio waves.
3. Give two reasons why "radar" with light waves would not be a very useful probe of distant planets.

4. List the sorts of information about a planet that could be obtained by radar and which would be difficult to obtain by other means. Name one type of information not readily obtained by radar.

5. Imagine a planet with a small axial inclination which has a *retrograde* sidereal orbital period of 100 days, and a sidereal axial period of -100 days. Discuss whether the planet is in synchronous rotation around the Sun. What is its *solar* axial period?

6. List *two* quantities that you would like to know, or to know better, plus *two* types of observation you would like to make, or to make better, in order to extend our knowledge of the Cytherean interior and of the level of geological activity past and present.

7. Outline *four* ways in which Venus is *known* to resemble the Earth, and *four* ways in which it is known to be strikingly different. What are the possible causes of the differences? List also *two* important ways in which it is not yet clear whether Venus and the Earth are appreciably different from each other.

8. Describe the three types of region recognized on the Cytherean surface.

9. Why are the seasonal changes so small on Venus, and why is the polar climate so similar to the equatorial climate?

10. Outline some of the evidence from which it is believed that water has not been lost in *large* quantities from Venus, and the evidence which indicates that *some* may have been lost.

5

THE ACQUISITION OF VOLATILES BY THE TERRESTRIAL PLANETS

Though volatiles represent only a *tiny* fraction of the masses of the terrestrial planets, they endow Venus, the Earth, and Mars with interesting phenomena such as atmospheres, oceans, polar caps and, at least in the case of the Earth, a biosphere. The Moon and Mercury are far less well endowed with volatiles as you will see in subsequent chapters.

There are various ways in which the terrestrial planets, including Mercury and the Moon, could have acquired their volatiles.

First, it is possible that the volatiles were borne on the small bodies of dispersed material from which the planets accreted. All of these small bodies could have carried comparable amounts of similar volatiles as indicated in Figure 5.1 (a) by the uniform distribution of spots in the planet, so that in this sense the accretion was homogeneous. Alternatively the small bodies acquired by a planet at different stages of accretion could have borne different amounts or different types of volatiles. In this sense the accretion was heterogeneous: in Figure 5.1 (b) is shown the case of volatile-rich materials that have been added on top of less rich materials.

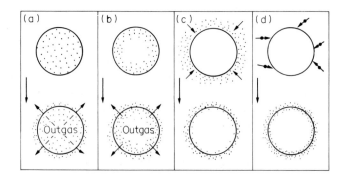

FIGURE 5.1
Different ways of acquiring volatiles.

Second, it is possible that as the mass of the planet built up the volatiles were captured *as gases* from the planetary formation medium (PFM) by the gravitational field of the (growing) planet. This is illustrated in Figure 5.1 (c).

Third, it is possible that the volatiles were acquired through the impacts on otherwise completed planets of volatile-rich bodies, as illustrated in Figure 5.1 (d). In a sense this is an extreme form of heteregenous accretion of volatiles, though the separation of planetary accretion from the veneering by volatiles is so complete that it merits separate consideration.

If volatiles are acquired during accretion then most of the volatiles are initially buried deep in the planet. And they will remain there unless the interior becomes hot enough for the volatiles to escape. The most difficult volatiles to dislodge will be those borne in compounds such as carbonates and hydroxy silicates.

The transfer of volatiles from the interior of a planet to its surface is called *outgassing*. In the case of the Earth outgassing occurs today via volcanoes, and also via *fumeroles* which are holes in the ground which emit volatiles but which emit little ash or lava, and therefore, unlike volcanoes, do not build mountains. There are also less dramatic seepages of volatiles on to the Earth's surface. It certainly seems likely that the interior of the Earth has been hot enough for most of any deeply buried volatiles to have found their way to the surface, particularly if convection extends throughout the mantle. But the present existence of outgassing from the Earth's interior does *not* prove that the Earth's volatiles were initially deeply buried. It is clear that most of the volatiles being outgassed today have been outgassed before: they are being recycled by geological processes, particularly by the subduction of plates. It is possible that nearly all of the outgassed volatiles have always been the result of recycling.

In earlier chapters the possibility has been raised that the atmosphere of the Earth, and perhaps of Mars and Venus, originally contained CH_4 and NH_3 rather than CO_2 and N_2. It is plausible that the internal temperatures and chemical conditions were then such that CH_4 and NH_3 was outgassed rather than the CO_2 and N_2 which outgases today. It is also plausible that such planets were veneered by CH_4 and NH_3. In either case the atmosphere would subsequently revert to CO_2 and N_2, as outlined in section 2.4.

Clearly therefore, the present outgassing of the Earth, and the possibility of an early atmosphere containing CH_4 and NH_3, provide no firm bases on which to decide between the various ways in which the terrestrial planets could have acquired their volatiles.

5.1 The evidence of the inert gases

A source of evidence that is often brought to bear on this question is the relative abundances of certain inert gas isotopes. The *inert gases* are so called because they do not readily form chemical compounds. They are also *gases* at all but extremely low temperatures. The complete family comprises helium (He), neon (Ne), argon (Ar), krypton (Kr) (from which the authors of *Superman* got the name of his home planet), xenon (Xe) (pronounced zenon) and radon (Rn). Certain of the various inert gas isotopes have to be excluded, some because they are of sufficiently low mass to escape in appreciable quantities from the terrestrial planets, others because they are produced by the radioactive decay of isotopes of elements other than inert gases, and others for both reasons. Some are omitted on grounds of scarcity.

The survivors are shown in Figure 5.2 (along with certain isotopes of H, C and N to which I shall turn in the next section). Figure 5.2 shows for the Earth, Venus and Mars the abundances of the inert gas isotopes as the number of *atoms in the atmosphere* per *kilogramme* of the whole planet. This leads to very large numbers, but don't be misled into thinking that the inert gases are extraordinarily abundant: a kilogramme of a typical solid contains about 10^{25} atoms. Note that in Figure 5.2 a *logarithmic* scale of abundance has been used, in which equal distances correspond to equal multiples. This is to accommodate the large range of abundances. Note also that the abundance of ^{132}Xe has not yet been measured for Venus.

Figure 5.2 also includes the solar relative abundances of the same set of isotopes, and these are taken to represent the PFM. The mass of the PFM is unknown and therefore the abundances are *not* to be regarded as the number of atoms per kilogramme of PFM, ^{20}Ne being arbitrarily placed at 10^{22} atoms per kilogramme to facilitate comparison with the graphs for the planets.

It is at once clear from Figure 5.2 that the PFM graph is a different shape

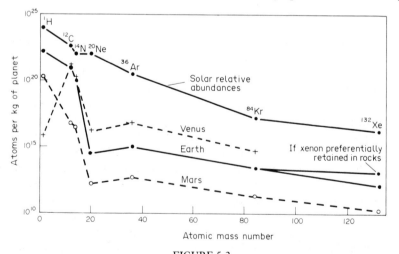

FIGURE 5.2
Planetary abundances of inert gases and of some elements which form only volatile compounds.

from those of the three planets, which are broadly similar to each other. The capture of inert gases from the PFM cannot by any plausible means lead to this discrepancy and therefore this is ruled out as the major mechanism by which these planets acquired these inert gases.

You might imagine that somehow the discrepancy in Figure 5.2 can be explained by supposing that the planets have withheld from the atmosphere different proportions of their supply of the different isotopes. However, the chemical inertness of the inert gases means that though they can be physically trapped in rocks they cannot be chemically combined to form significant quantities of non-volatile materials. Physical trapping is likely to be about as effective for neon, argon and krypton. It may be more effective for xenon, which has a comparatively large and massive atom and would therefore leak out less readily than the others. In the case of the Earth it has been estimated that this could lead to an underabundance of xenon in the atmosphere of a factor of 10 or more. However, as shown in Figure 5.2, this makes the discrepancy *worse*.

If the gas density in the PFM was appreciable then the terrestrial planets could have acquired substantial atmospheres, and these would have been representative of the PFM, with H_2 and He dominant. As the PFM dissipated then H_2 and He alone would largely escape. However, Figure 5.2 indicates that the present inert gases are not the surviving remnants of such an atmosphere, and therefore these planets have lost any such atmospheres.

This brings me to volatiles borne by small bodies, either during accretion, or subsequently as a veneer.

Small bodies, like planets, can acquire *small* quantities of volatiles by the direct capture of gases from the PFM, which are then adsorbed on to their surfaces. The surface area of N small bodies is roughly the cube root of N larger than a single body formed by their accretion. In the case of a planet N is *very* large and therefore a considerably larger quantity of volatiles can be acquired by the planet by this means than it could acquire by capturing gases directly from the PFM.

The small bodies will have themselves accreted from even smaller bodies, the surfaces of which would carry traces of volatiles. This can lead to volatiles trapped *inside* the small bodies, that is, *occluded* in them. Moreover, some of the small bodies that reach the Earth could come from regions of the PFM far colder than the region where the Earth formed. In such regions the temperatures could have been low enough to permit certain volatiles to condense yielding volatile-rich bodies which could survive passage to the Earth. The abundant volatiles particularly prone to such condensation are H_2O, CO_2, CH_4 and NH_3, that is, the common icy materials.

Small bodies can also bear volatiles in less volatile compounds, such as carbonates from which CO_2 can be obtained, and hydroxy silicates from which water can be obtained.

Through all these means small bodies can bring to the terrestrial planets volatiles in sufficient abundance to match the observations.

It is sometimes argued that if the Earth, Venus and Mars were veneered by volatile-rich bodies *subsequent* to the main phase of accretion, then the inert gas graphs for the three planets in Figure 5.2 should lie roughly on top of each other. That this is not so, it is argued, shows that the volatiles must have been acquired during accretion. The argument rests on the assumption that the number of volatile-rich bodies that would be captured by a planet after it has accreted is roughly in proportion to the mass of the planet. However, this proportionality with mass need not hold. Moreover, the argument is insecure on other grounds. Thus, it could be that because the three planets occupy different parts of the PFM they captured somewhat separate populations of volatile-rich bodies. Also, conditions inside the three planets could be sufficiently different that different proportions of their volatile endowments have come to be withheld from the atmospheres. Moreover, if the planetary volatiles were acquired *during* accretion then it is still necessary to explain the displacements of these graphs in Figure 5.2. There *are* plausible ways of doing this, but this does not rule out post-accretional veneering.

Another argument ranged against post-accretional veneering is based on the inert gases in meteorites, particularly the carbonaceous chondrites (section 14.5) which many scientists believe typify the small bodies from which much of the Earth accreted. This is plausible but by no means necessarily the case, and therefore the evidence from meteorites has to be treated with considerable reserve. However, were the graph for the inert gases in carbonaceous chondrites to be plotted in Figure 5.2 it would have the same shape as that of the planets. This can be regarded as support for the view that such meteorites typify much of the material from which the three planets accreted. More convincingly, it shows that the acquisition of inert gases by small bodies can yield a different graph from that of the PFM. But when the graph is used to argue against post-accretional veneering certain difficulties arise.

A major difficulty is that the inert gases account for only a small fraction of the volatiles known at the surfaces and in the atmospheres of the three planets, where the volatiles almost entirely consist of CO_2, H_2O, N_2 and O_2, or of compounds derived from them such as carbonates, hydroxy-silicates and nitrates. The relative abundances of such volatiles in meteorites does *not* match those of the three planets. This is hardly surprising, because the chemical *non*-inertness of the dominant volatiles and the considerable differences in volatility between one volatile and another and between their compounds leads to numerous possible ways in which differences could arise. However, this leaves open the possibility that the Earth, Venus and Mars acquired most of their volatiles from volatile-rich bodies by post-accretional veneering.

Thus, whilst the similarity of the inert-gas graph for meteorites to those of the planets in Figure 5.2 certainly supports the view that those *inert gases* were acquired *during* accretion, it remains *possible* that the *dominant volatiles* were acquired by the capture of volatile-rich bodies *after* accretion.

5.2 The evidence of hydrogen, carbon and nitrogen

Figure 5.2 also displays the abundances of the most abundant isotopes of hydrogen, carbon and nitrogen. These are constituents of the abundant volatiles H_2O, CO_2, N_2. Oxygen has been omitted because nearly all the oxygen in a planet resides in non-volatile oxides and silicates and it is extremely difficult to estimate how much of this oxygen has been liberated into more volatile forms and conversely how much oxygen originally in volatile forms has been lost by combination with various oxides and silicates. Sulphur, a component of the volatile SO_2, is excluded for the same sort of reason. By contrast hydrogen, carbon and nitrogen are all associated with volatile compounds or with compounds such as hydroxy silicates, carbonates and nitrates that yield up volatiles relatively readily.

For all three planets the atmospheric quantities of H, C and N are included, where these elements mainly reside in H_2O, CO_2 and N_2 respectively. In the case of Mars the atmospheric quantity for N has been multiplied by 10 to allow for the escape of N to space. In the case of the Earth the known surface repositories have been included, mainly the oceans for H, carbonates for C and nitrates for N. Surface repositories have *not* been included for Venus and Mars.

Figure 5.2 shows that on all three planets the relative abundances of carbon and nitrogen with respect to neon (and therefore with respect to the other inert gases) are far greater than in the PFM — remember the logarithmic scale. This indicates that the inert gases were acquired by a different means from carbon and nitrogen. Moreover, this possibility is *strengthened* by the neglect of surface repositories, because regardless of how the various volatiles were acquired it is likely that a smaller fraction of the inert gases are retained in such reservoirs than are carbon and nitrogen. Therefore this is evidence for the view that whereas certain volatiles such as the inert gases were acquired *during* accretion, others such as carbon and nitrogen (and hydrogen) were acquired subsequently. However, planetary chemistry is sufficiently complicated that this cannot be regarded as a firm conclusion.

Figure 5.2 can also be used to make comparisons *between* the three planets. You can see that Mars is depleted in H, C and N with respect to the Earth by factors of between about 100 and 1000. Were all the surface repositories on Mars to be included then the depletions would be reduced, but it is not certain that they would be eliminated. This possible depletion can be explained in several ways. First, *if* Mars acquired volatiles during accretion then the interior could be less thoroughly outgassed than that of the Earth, and this is in accord with the evidence for the lower level of geological activity on Mars. Second, Mars could be fairly thoroughly outgassed but accreted from materials less well endowed with volatiles than the Earth. Third, Mars could have been less well veneered with volatiles after accretion was complete. It is not known which of those possibilities represents the truth.

In the case of Venus you can see from Figure 5.2 that the abundances of C and N are fairly similar to those on the Earth, and it is possible that this is true of the planets as a whole. However, the deficiency of H is clear, and there are no plausible means by which H can be retained in the interior in substantially greater quantities than C and N. In section 4.4.3 certain difficulties were noted of Venus losing to space large quantities of H. To these difficulties must now be added the difficulty that *if* Venus has lost large quantities of H to space then the Earth, with its warm exosphere and Mars with its low gravitational field, should also have lost large quantities of H. This supports the conclusion that Venus has always been deficient in H, but it does not help explain *why*.

The volatile endowments of the Earth, Venus and Mars will be further discussed in Chapter 15.

5.3 Mercury and the Moon

The surfaces of the Moon and Mercury are *extremely* depleted in volatiles compared to the Earth, Venus and Mars. The lunar surface is also depleted in *moderately* volatile substances, and the same is probably true of Mercury.

There are four possible reasons for these depletions.

First, Mercury and the Moon could have accreted from materials that were far less well endowed with volatiles than were those from which the Earth, Venus and Mars accreted. This explanation relates to the acquisition of volatiles *during* accretion.

Second, if volatiles were acquired *after* accretion, from the impacts of volatile-rich bodies, then Mercury and the Moon could have failed to capture the volatiles made available in this way. The extent to which an atmosphere-less planet can hold on to such volatiles depends on the impact speed of the volatile-rich body and on the gravitational field of the planet. It is plausible that Mercury and the Moon allowed the volatiles to "rebound" into space.

Third, Mercury and the Moon could have allowed nearly all of their original volatile endowments to escape to space. The rate of thermal escape depends on the temperature of the exosphere and on the gravitational field. The rate of chemical escape also depends on the gravitational field and also on the details of chemical reactions in the upper atmosphere. Though it is plausible that Mercury and the Moon could have lost large quantities of volatiles to space it seems unlikely that this explains the extreme extent of their depletions.

Finally, Mercury and the Moon may be much less thoroughly outgassed than the other terrestrial planets. However, it is hard to see how their surface regions could be so extremely depleted in volatiles if the interiors are not significantly depleted.

5.4 Summary

It is extremely difficult to determine how Venus, the Earth and Mars acquired their volatiles. However, it is very likely that capture of *gases* from the PFM by a largely formed planet can be ruled out as a major source.

Mars, compared to the Earth, may be depleted in H, C and N, and it seems fairly certain that Venus is heavily depleted in H. The surfaces of Mercury and the Moon are *very* heavily depleted in all volatiles, and in the main this is probably because of a smaller initial endowment.

5.5 Questions

1. In Figure 5.2 you can see that the Earth has about 10^{21} atoms of ^{12}C per kilogramme of the Earth. If no repositories of ^{12}C have been overlooked then approximately what fraction of the atoms of the *whole* Earth are ^{12}C?
2. List the different ways in which a planet can acquire and lose its volatiles, and in each case indicate the degree of outgassing required to concentrate the volatiles at or near the surface.
3. Why is the evidence from the inert gases, and from H, C and N, inconclusive with regard to how the Earth, Venus and Mars acquired volatiles?
4. Outline four possible reasons for the near absence of volatiles on Mercury and the Moon.

6

THE MOON

The Moon orbits the Earth and is sufficiently close that we can see its disc with the unaided eye. The features of the "Man in the Moon", recognized by many cultures, consist of dark and relatively smooth areas called maria (singular "mare", pronounced "ma-ray") lying amidst bright rugged areas called highlands. "Mare" is Latin for sea, because when this appendage was given in the seventeenth century it was believed that this is what they were. However, it is now known that the Moon is devoid of seas, and devoid of an atmosphere as well. It has been extensively explored by spacecraft and is the only extraterrestrial body on which men have stood. The first manned landing was by Apollo 11 in July 1969.

6.1 The Earth and the Moon

The Moon is in synchronous rotation around the Earth, and therefore within the limitations imposed by its orbit always shows much the same face to the Earth. Figure 6.1 (a) shows this familiar face, and Figure 6.1 (b) shows the far side which was revealed by spacecraft. The most abundant landform on both sides is the impact crater, the larger of which are visible in Figure 6.1. You can also see that the maria are largely confined to the near side.

When the Moon moves between the Earth and the Sun it is seen from the Earth to move across the Sun, and this is called a *solar eclipse*. The Moon's angular diameter, as seen from the Earth, is comparable to that of the Sun, and therefore in many solar eclipses, from a rather narrow region across the Earth's surface, the Sun will be totally covered by the Moon. This is a *total solar eclipse*. The narrowness of the region, and its variable position from one eclipse to the next, means that at a given point on the Earth's surface total solar eclipses are rare. In 1973 I was fortunate enough to see a spectacular total solar eclipse at Lake Turkana in northern Kenya, with the bonus that this wild and exotic region is worth visiting in its own right.

Largely because the Moon is acted on to a significant extent by both the Sun *and* the Earth its orbital elements are subject to relatively rapid changes. Consequently the dates of total solar eclipses form a rather complex sequence. Those occurring in the near future are listed in Table 1.2.

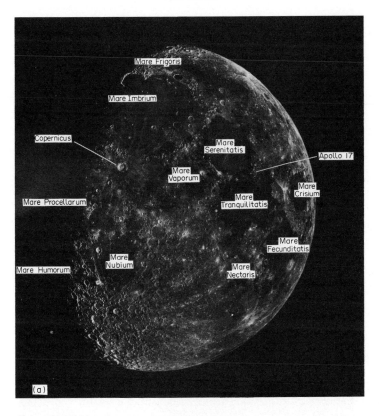

FIGURE 6.1
Both sides of the Moon. (a) the familiar side imaged from the Earth, (b) the far
side imaged by Lunar Orbiter 4 with Mare Orientale at the centre. North is at the
top in both cases.

The synchronous rotation of the Moon around the Earth is one of several
consequences of tidal interactions between the Earth and the Moon, ocean
tides being a more familiar example. Tidal interactions are important
throughout the Solar System.

● 6.1.1 Tidal interactions

Consider two bodies in orbit around each other, such as M and m in Figure
6.2 (a). (As outlined in section 1.1, they will orbit around the centre of mass,
the position of which depends on the masses M and m.) The gravitational
attraction between them will be strongest between their closest points, a and
a', and weakest between their furthest points, c and c'. A force which, as in

(b)

FIGURE 6.1
(continued)

this case, varies across a body is called a *differential force*, and a differential gravitational force is called a tidal force.

The tidal forces between M and m tend to distort the bodies from their spherical shapes, producing elongations along the line joining the two bodies as shown in Figure 6.2 (a). These tidal forces are opposed by the internal gravitational forces which tend to make bodies spherical, and by non-gravitational intermolecular forces which resist the separation of molecules. If the tidal forces are not too powerful then an equilibrium will be reached in which there is a certain degree of distortion. This distortion is called a *tide*.

(If, as on the Earth, there are outer "shells" of oceans and atmosphere then tides will appear in these too and they will generally be greater than in the rest of the planet thus producing variations in oceanic and atmospheric thickness around the planet.)

Suppose that the two bodies in Figure 6.2 (a) are *not* in synchronous rotation with respect to each other. It follows that the tides tend to get carried off the line joining the two bodies, and they can only keep on line if the tides move like slow waves across the surface of each body. In practise these *tidal waves* cannot quite maintain the alignment, resulting in the tides making a small and roughly constant angle with respect to the line joining the two bodies.

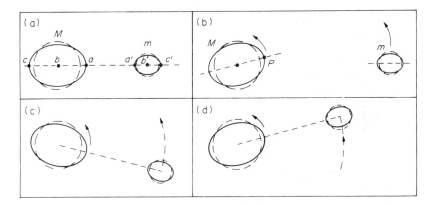

FIGURE 6.2
The influence of tidal forces on planetary rotation.

A particular case is illustrated in Figure 6.2 (b). The axial spin of the more massive body M tends to carry the tide *ahead* of the orbital motion. The less massive body m is in synchronous rotation and the orbital eccentricity is small, and therefore the tide on m always lies along the line to M. Because the tide on M is misaligned then m can exert a twisting force on M called a *torque* and this tends to produce alignment. This torque will gradually reduce the axial spin rate of M until it too is in synchronous rotation.

Had the axial spin of M been *slower* than that of synchronous rotation then the torque would have acted to speed it up until synchronous rotation was achieved.

Thus, tidal forces can lead to synchronous rotation. It can be assumed that m in Figure 6.2 (b) has acquired its synchronous rotation through the operation of such forces. Such forces also reduce axial inclinations.

If the orbits of M and m around each other are not circular then, with or without synchronous rotation, during each orbit the tides on each body will oscillate around the average orientation, the axial rotation rates being constant. This effect can be shown to reduce orbital eccentricities.

Tidal forces can also change the semi-major axis of an orbit. From the principle of conservation of angular momentum, which you shall meet in section 6.5.1, it can be shown that if the combined axial spin of the two bodies is *decreasing* then the semi-major axis must be *increasing*, and vice versa. For example, in Figure 6.2 (b) the axial spin of *M* is slowing down whereas that of *m* is constant. Therefore, the two bodies must be moving further apart. Moreover, as their separation increases, tidal forces will also cause *m* to slow down to maintain its synchronous rotation. This increases the rate of separation.

Whether *m* and *M* move apart depends on the semi-major axis of the orbit in relation to that of the *orbit of Keplerian co-rotation*. This is the orbit in which the less massive body *m* would have a sidereal *orbital* period equal to the sidereal *axial* period of *M*, as illustrated in Figure 6.2 (c) and (d). It can be shown that if *m* lies *outside* this co-rotation orbit then ultimately *M* and *m* will move apart, but if *m* lies *inside* this co-rotation orbit then ultimately *M* and *m* will move nearer. A simple case occurs when *M* is far more massive than *m*, in which case *m* has little effect on the axial rotation of *M* and changes in the semi-major axis of the co-rotation orbit are negligible.

Tidal forces can also heat a body, because the movement of a tidal wave through a body means that the material of the body is being flexed, and this will generate heat. The energy for *tidal heating* is drawn from the reduction in rotational energy consequent upon the various changes in axial and orbital motion.

Tidal forces can also tear bodies apart. The closer two bodies are together the greater the tidal forces between them. There will be a limiting distance within which the more massive of the two bodies will disrupt the other, and a smaller limiting distance within which, had it survived, the less massive body would have disrupted the more massive body. This is called *tidal disruption*.

People and spacecraft are far too strongly bound by *non*-gravitational forces to be tidally disrupted by any bodies in the Solar System. However, the larger the body the less important become its non-gravitational forces in comparison with its internal gravitational forces. This is because the gravitational forces are appreciable over much larger distances than the non-gravitational forces. Ultimately there is a direct contest between internal gravity and the tidal forces. The minimum radius of the body for which this direct contest is established depends on the *density* of the body. For bodies consisting of rocky or icy materials the minimum radius is about 1000 km. In such a direct contest the limiting distance within which the more massive body disrupts the other is called the *Roche-limit* after Edouard Albert Roche (1820–1883), the French scientist.

The Roche-limit depends on the *mass* of the more massive body and on the *density* of the less massive body: the greater the mass the smaller the limit, and the greater the density the smaller the limit.

6.1.2 *Tidal interactions between the Earth and the Moon*

Figure 6.2 (b) can be taken to represent the Earth and the Moon today, though it is not to scale and the tides have been greatly exaggerated. Only the Moon has acquired synchronous rotation because the tidal forces of the Earth on the Moon are greater than those of the Moon on the Earth. However, the residual eccentricity of the lunar orbit, plus other effects, means that we can see over half, about 59%, of the lunar surface from the Earth.

The axial rotation rate of the Earth is directly observed to be slowing down, such that the sidereal day is lengthening by about 1.5×10^{-5} second per year. Moreover, from the annual and diurnal growth patterns of organisms, preserved as fossils, it has been established, for example, that about 380 Ma ago there were about 400 "days" in the year and as the orbital period of the Earth is very unlikely to have changed the day must then have contained only about 22 hours.

Far into the future the Earth and the Moon will both be in synchronous rotation about each other, the sidereal orbital and axial periods all being about 55 days. The orbit will be circular and the Moon will be about 1½ times as far away as it is now and so total solar eclipses will not occur. The axes of the Moon and the Earth will be perpendicular to their orbit around each other, and because this orbit will be less inclined to the ecliptic than is the Earth's axis today, seasonal variations on the Earth will be less.

It is just possible that this far-off configuration will be unstable, and that the Sun will draw the Moon away into an independent orbit around the Sun, leaving the Earth Moon-less.

6.2 The interior of the Moon

6.2.1 *The evidence of gravity*

The mass of the Moon has been obtained from the position of the centre of mass of the Earth and the Moon which lies 4748 km from the centre of mass of the Earth. This position was long ago determined from the slight motion of the Earth around this point. This gives the mass of the Moon as a fraction of the mass of the Earth, and the mass of the Earth is known. The mass of the Moon has also been determined from spacecraft paths in the lunar vicinity, and from the Doppler-shifts in the wavelength of spacecraft radio transmissions to the Earth, shifts induced by the motion of the Earth around the Earth–Moon centre of mass.

The Moon's mass is $0.01230 \, M_E$ and its mean radius is 1737 km. Therefore, its mean density is 3340 kg/m^3. Because of the small mass of the Moon the uncompressed density will not be much less. The uncompressed density of the Earth is considerably greater than that of the Moon and this strongly implies that the Moon is strongly depleted in metallic iron compared to the Earth.

The polar moment of inertia of the Moon C has been determined from J_2, and from the *physical librations* which for the Moon have played an equivalent role in determining C to that played by the rate of polar axis precession in the case of the Earth (see section 3.2.1). The physical librations arise because the path of the Moon around the Earth is not perfectly circular. As a consequence the tide *in the Moon* does not always lie along the line from the Moon to the Earth. This enables the Earth to exert a torque on the Moon which causes small but observable oscillations. These are the physical librations. From these librations, and from J_2 which is obtained from the orbits of spacecraft, C can be calculated.

For the Moon C is not very different from that of a homogeneous sphere and indicates a very slight overall increase of density with depth. However, because of the small mass of the Moon the compression inside it is also small, and the slight increase of density with depth is *more* than can be explained by the compression of a homogeneous material. The Moon is *not* homogeneous.

The way of obtaining C outlined above does *not* assume that the Moon is in hydrostatic equilibrium, which is just as well because it is not. Moreover, its observed non-hydrostatic mass distribution *cannot* readily be understood as a hydrostatic state "frozen" from an earlier environment, such as when it was closer to the Earth. This suggests that the Moon, or at least the outer 200 km or so, has never been flexible enough to reach hydrostatic equilibrium and therefore that these outer regions have always been cold. However, it is also possible that the outer 200 km or so cooled and became rigid whilst they were deformed by internal turmoil which destroyed hydrostatic equilibrium.

Further evidence for a rigid outer Moon comes from the 5 km offest (towards the far side) of the centre of figure from the centre of mass. This offset has been determined from spacecraft orbits which locate the centre of mass of the Moon, and from the measurement of the altitude of a spacecraft above each point on the lunar surface over which it passes by timing the journey down and back of light pulses from lasers on board (compare section 4.3).

6.2.2 Seismic evidence

The astronauts of Apollos 11, 12, 14, 15 and 16 placed seismic stations on the Moon. Seismic data were returned to Earth over a period of a few years from these five stations, though they were reduced to four when the Apollo 11 station stopped working. These data are insufficient to establish the variation of seismic wave speeds with depth in the Moon to anything like the same extent with which this has been possible in the Earth. Figure 6.3 is about the best that can be done with the lunar data at present.

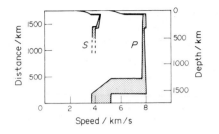

FIGURE 6.3
P-wave and S-wave speeds in the Moon.

At the Apollo 17 site an experiment was carried out by the astronauts which gave the seismic wave speeds down to a few kilometres depth in the vicinity of the site. Some evidence for a layered structure was found, but whether this is anything more than a local feature is unknown.

In Figure 6.3 the rapid increase of speed with depth in the outer 10 or 20 km is due largely to the increasing compaction of a fragmented medium as the overburden increases: it is not possible for solid rock to give this behaviour. Moreover, the lunar seismic signals were of a very different nature from those seen in the Earth, and this can be explained by the prevalence of a fragmented medium in the outer regions of the Moon. The signals also indicate that these regions are free of gases and liquids.

There are several other sources of evidence for the fragmented nature of the outer Moon.

At all sites at which there have been landings, manned or otherwise, the lunar surface is almost entirely blanketed in fine dust, with a few stones and boulders here and there. The middle distance of Figure 6.4 is typical. Moreover, Earth-based photometric studies have indicated that most of the lunar near side must be blanketed in such dust. Direct sampling of the dust at various landing sites has indicated that it extends downwards for at least a metre or so. Photometry of the Moon with Earth-based radar shows that over much of the near side the dust extends downwards for at least about a hundred metres. Analysis of the radio waves emitted by the Moon indicate that the thermal conductivity of the outer Moon increases rapidly with depth, just as would be expected from the increasing compaction of a fragmented medium.

However, it remains an open question whether the dust continues beyond about 100 metres, becoming fully compacted rock at 10 to 20 km, or whether it gives way to broken rock somewhere not far beyond 100 metres depth. Certainly, little fragmentation could exist beyond about 20 km because of the high pressure of the overburden. Beyond this depth the compression of solid rock such as silicates can account for the increasing wave speeds.

FIGURE 6.4
The lunar surface at the Apollo 17 site, imaged by the Apollo 17 astronauts.

At about 50 km depth, as you can see from Figure 6.3, the wave speeds suddenly increase. This could result from a *structural* change in a single solid material, though no suitable materials are known. It could also result from a change in composition, and thus indicates that the Moon has a crust. However, because all the seismic stations have been placed on the near side the wave speeds down to this depth only reliably apply to the near side. There may be no crust on the far side, and even on the near side it *may* be confined to certain regions, such as beneath the maria, though the majority view is that the Moon *does* have a global crust.

If there is a lunar-wide crust then *one* explanation of the offset of the centre of mass from the centre of figure is that the crust on the far side is, on average, thicker than on the near side. Plausible values are 75 km and 50 km respectively.

At depths greater than about 100 km the wave speeds are rather uncertain. Figure 6.3 is not a unique analysis, but does correspond to a medium consisting mainly of silicate materials.

You can see that there may be a compositional change at about 300 km depth, and there *may* be a core about 170 to 450 km radius.

You can also see that the S-wave speed decreases from about 50 km depth, and that beyond about 800 km it may not be possible for S waves to traverse the Moon. These observations can be explained by a modest increase of temperature with depth, the material becoming very plastic, perhaps partially molten, beyond about 800 km.

There are three sorts of seismic wave sources on the Moon. First, there are surface impacts from space vehicles and from meteorites. Second, there are interior sources which have only "fired" once during the several years of observations. Third, there are a few sources which fire repeatedly: these lie at depths between about 300 km and 800 km and may be the result of tidal stresses produced in the Moon by the Earth.

Most seismic sources in the Earth lie no deeper than about 100 km and generate *much* more seismic energy than all the lunar sources put together. They are largely the result of geological activity. Clearly the "seismic quietness" of the Moon indicates a low level of geological activity.

6.2.3 Other evidence about the lunar interior

Spacecraft observations have placed an upper limit on the internal magnetic dipole moment of the Moon of 10^{-7} of that of the Earth. Thus, *if* the Moon today has a small liquid iron core then *either* its small size *or* the lack of a solid inner core *or* the slow axial spin of the Moon (sidereal period 27.32 days) has prevented it generating an appreciable magnetic field.

Weak magnetic fields have been detected in surface rocks. In some cases these seem to have been produced by surface impacts, but in other cases exposure to a rather strong magnetic field in the past seems to be required. A possible source of such a field is an internal lunar dipole source, since vanished, though other possible sources are the Earth's field and the Sun's field if they were a good deal stronger in the past than they are today.

Some indication of the internal temperatures in the Moon has been gleaned from a weak magnetic field that is generated by the *solar wind* (see section 9.1.1). Because the Moon has no atmosphere nor any significant magnetic field of its own to deflect the solar wind, it sweeps near the lunar surface. The solar wind carried its own magnetic field, and as this sweeps past it causes electrical currents to flow in the Moon. These currents generate their own magnetic field which can be distinguished from the various other fields. (Indeed, were the fields in the solar wind much stronger in the past then this is another way in which the magnetic fields that reside in the surface rocks could have been produced.) The magnetic field generated in the Moon by the action of the solar wind enables the electrical currents flowing in the Moon to be deduced, and this enables rough estimates to be made of the internal temperatures, the higher the temperatures the greater the currents for a given solar wind field. The temperature seems to increase with depth, reaching about 1000 K at about 300 km and about 2000 K at about 1000 km. It is not known whether the temperature continues to increase towards the centre.

These temperatures are consistent with the apparent rigidity of the outer 200 km or so of the Moon and with the strong attenuation of S waves beyond about 800 km. They are also consistent with the small size of the Moon.

The rate at which heat is flowing out of the Moon was measured at the Apollo 15 and 17 landing sites, but because of the unknown distribution of energy sources in the Moon, the poorly known thermal conductivity of subsurface materials, and the likelihood that the Apollo 15 and 17 results do not typify the whole Moon, the heat-flow measurements do not much constrain models of the lunar interior.

6.2.4 Models of the lunar interior

Figure 6.5 is a model of the lunar interior, many features of which would be accepted by most scientists, though a minority would argue that it has yet to be established that the Moon has a global crust.

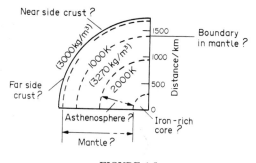

FIGURE 6.5
A model of the Moon.

Some modification of this model is certainly possible. For example, the average density of any crust could be increased and the density or size of any core could be correspondingly decreased within fairly broad limits. Also, any core may consist largely of iron, or of iron-rich compounds such as iron sulphide (FeS) and various iron oxides (FeO, Fe_2O_3), or of some mixture of iron and iron-rich compounds.

However, it seems fairly certain that the Moon is depleted in iron with respect to the Earth, not only in iron alone but also in iron-rich compounds because iron and these compounds are too dense to be sufficiently abundant to make up the iron abundance to that of the Earth. It also seems fairly certain, from the seismic wave speeds and from the densities based on the polar moment of inertia, that though the outer 300 km or so very probably consist largely of common silicates this detailed composition is not the same as in the Earth's mantle.

6.3 The lunar surface

The most common feature on the lunar surface is the impact crater. And yet until the Space Age it was not at all certain whether the lunar craters were of impact origin or of volcanic origin. But the close scrutiny of the lunar craters and examination of lunar samples plus theoretical investigations and laboratory experiments have established beyond reasonable doubt that all, or very nearly all of the lunar craters are of impact origin. Such craters are now known elsewhere in the Solar System as you have already seen in the case of Mars, and therefore the mechanism by which they are formed is of widespread relevance.

● *6.3.1 The formation of impact craters and associated features*

When a sizeable body from space strikes a planet's surface the impact speed will *not* be less than a few km/s. Therefore the impact is very violent. Most, or all of the impacting body will vaporize, and the resulting bubble of gas will expand with explosive violence, much more rapidly than the speed of the body during its final stages of decceleration. Therefore, the outward motion of the expanding bubble is not significantly biased in the direction of the impact, and a roughly *circular* crater is excavated. Only when the angle of impact is less than about 20° above the surface is the direction of the body preserved, as an elongation of the crater in the impact direction.

Material is thrown out by the impact in various forms, collectively called *ejecta*. "Bounce back" of the compressed planetary surface aids the ejection of material. Among the ejecta there will be vaporized material, some from the impacting body, but largely from the surface of the planet. This will condense and form a fine dust much of which will return to the surface, most of it a long way from the impact. Another far-flung component will be droplets of liquefied rock. These will rapidly solidify to form small beads of *glass*, a type of solid with a disorderly molecular arrangement. The impact will also generate *brecchias* (bretchi-ars). These are rocks formed by the welding of rock fragments by heat and pressure, producing a rock in which there is little empty space but in which the original fragments can be discerned. Conversely, impacts will also shatter rocks into small fragments. Some of the brecchias and rock fragments will also be scattered far away.

Other forms of ejecta lie fairly close to the impact, and unlike the far-flung material can normally be associated with a specific impact. Among this material there will be the ejecta blanket (section 3.3.5). There may also be bright *rays*, roughly resembling the spokes of a wheel radiating from the crater "hub". Such rays consist of bright subsurface material. There are also likely to be *secondary craters*, caused by large blocks of material flung out by the primary impact. Secondaries can usually be identified because they cluster around the larger primary, often forming chains and other small groups, and because the impacts producing them are at low angles and low speeds and so the craters are usually elongated. There may also be signs that liquid rock flowed around the crater, from impact melted materials.

The ejecta volume exceeds the volume of the imapcting body, usually by a large factor, and because little of these ejecta fall back into the crater the crater will also be larger than the impacting body by a similar factor. It is the *energy* of the impacting body that largely determines the volume of the crater, the larger the energy the larger the volume. The energy rises as the mass and impact speed of the body rise. The composition of the impacting body has a smaller effect. The impact speed depends on the mass of the planet as well as on the orbit of the body. The resulting crater and its ejecta will also depend on the direction of the impact, the nature of the planet's surface, and the nature of any atmosphere.

The shape of a crater depends on its size. Usually it is only the smallest craters that have a simple bowl shape with the cross-section shown in Figure 6.6 (a), and craters less than 1 km diameter may even have no noticeable rim. Larger craters are less simple, as follows.

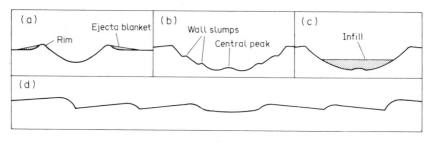

FIGURE 6.6
Cross-sections through different types of impact crater.

Figure 6.6 (b) shows a crater with a *central peak*. There may be several such peaks, and there are several ways in which they can be produced.

The largest craters of all, more than about 300 km in diameter, are usually called *impact basins*. A typical form is shown in Figure 6.6 (d), and you can see that there are several concentric rims, in which case it is called a *multi-ringed basin*. These rings could be the result of surface waves spreading out from the impact through the solid surface, or the result of a succession of wall slumps in a once deeper basin.

Wall slumps can also occur in smaller craters yielding the typical form in

Figure 6.6 (b). Such slumps could be triggered by seismic waves. They could also result from impact in a stratified surface.

A relationship has been established between crater diameter and crater depth, the greater the diameter the smaller the depth to diameter ratio. This trend is apparent in Figures 6.5 (a) and (b).

Floors in craters of any size can be flattened by *infill* as illustrated in Figure 6.6 (c). This infill could come from ejecta falling back into the crater, including molten rock from the impact, lava from volcanic activity, and dust from the walls and beyond. In applying the depth–diameter relationship allowance must be made for unusual degrees of infill.

In the larger craters there will be some upward movement of the crater floor towards isostatic equilibrium, material having been removed by the impact over a wide region. If there is subsequent substantial infill this could complete the isostatic adjustment or even produce a mass excess if the crater region has been compressed by the impact and is fairly rigid.

At the other end of the size range there is *microbombardment*, that is, bombardment by micron-sized particles. This does not produce craters but an erosive effect akin to that produced by wind-borne dust on the Earth.

6.3.2 Lunar impact craters

Figure 6.7 is a typical fairly large lunar crater, named Copernicus after the Polish astronomer Nicholas Copernicus (1473–1543). It is about 100 km diameter, and making reasonable assumptions about the impact speed would have been excavated by a rocky body about 10 to 20 km diameter. You can see several of the topographic features discussed in the preceding section: a floor lower than the terrain beyond the rim; the central peaks; the wall slumps; a floor flattened by infill; an (rather degraded) ejecta blanket; small surrounding craters, some of which can be shown to be secondaries. Elsewhere on the Moon other crater features can be seen. Thus, in Figure 6.1 (a) you can see crater rays and in Figure 6.1 (b) you can see a multi-ringed basin.

All of the various topographic features observed in lunar craters indicates their impact origin. Moreover, the diameter–depth relationship is as expected for impact craters on the Moon, and brecchias and glass particles are common. Only in a few cases is it unclear whether a crater is volcanic or from an impact, and in comparably few cases does it seem more likely that a crater is of volcanic origin. All these doubtful cases are *small* craters.

The lunar maria, nearly all of which are located on the near side, also seem to be of impact origin, with subsequent infill by rather dark material. I shall return to the nature of the infill in section 6.4.

There are several strong indications that the maria are the result of impacts. First, each mare is encircled, or nearly encircled by tall mountain

FIGURE 6.7
The lunar crater Copernicus, imaged obliquely by the Apollo 17 astronauts. The
crater is about 100 km across.

ranges that could be the remnants of the rim of a large impact basin or the
most prominent rim of a multi-ringed basin. Second, gravitational data show
that the maria are sites of excess mass called *mascons* such as could be
produced by infill of an impact basin as follows. It has been estimated that an
impact which would have excavated a basin the size of, for example, Mare
Imbrium (Figure 6.1 (a)) which is about 800 km diameter, would initially
have produced a bowl about 150 km deep. But even with a rigid outer Moon
gravitational forces would have raised the floor most of the way towards
isostatic equilibrium. However, even if such equilibrium was achieved the
floor would still lie below the surrounding terrain, because of permanent
compaction of the floor materials by the impact. It is estimated that the basin
would have become no deeper than about 5 km, and as today Mare Imbrium
is of negligible depth there could be up to about 5 km of infill. Such infill
could not only explain the mascon but could also explain why no Imbrium
impact features are visible within Mare Imbrium and why the surface is fairly
smooth. Moreover, on the far side of the Moon there are a few basins that, if

filled, would resemble mare. Mare Orientale on the far side (Figure 6.1 (b)) is partially filled, but displays a mass *deficit* (a "negative mascon") indicating that full isostatic compensation did not follow the impact.

The smaller the crater diameter the more numerous the craters, and this remains the case when secondary craters, which are smaller than their primaries, are excluded. This suggests that the smaller the mass the greater the number of impacting bodies. Bodies massive enough to excavate maria have been few in number though such bodies need not have exceeded about 100 km in diameter. It is not hard to find suitable populations of bodies that could give rise to the observed range of crater sizes today — more on this in section 6.4 and in Chapter 8. However, it is curious that the maria on the near side of the Moon significantly outnumber the maria and large basins on the far side — more on this in section 6.5.

6.3.3 Other surface features

Apart from impact craters, including mare basins, the most prominent lunar feature is the mare infill, about 17% of the lunar surface consisting of such infill. Most scientists believe the infill to be lava, and though there are no features yet seen on the Moon that are *definitely* volcanic, floods of lava from fissures that are buried beneath their own lava are known on the Earth and could account for the mare infill on the Moon.

However, it seems unlikely that the mare basins could each have been filled in one outpouring. Any such deep body of liquid is likely to rise to a level corresponding to isostatic equilibrium, and there would be none of the observed mascons. Moreover, the mare surfaces are slightly tilted, whereas a deep pool of liquid would solidify with a horizontal surface. Therefore, most of those scientists who believe that the infill is lava believe that it issued in a series of sheets, each a few tens of metres thick. This could yield mascons and tilted surfaces. The best evidence for such sheets comes from Mare Imbrium as shown in Figure 6.8. This exhibits a series of winding *scarps*, a scarp being a short steep slope that leads to flatter terrain. Each scarp could be the solidified flow front of a lava sheet, in which case the sheets were a few tens of metres thick. No such scarps have been observed on other maria, and it has been suggested that the lava there was of lower viscosity and would thus rapidly spread out into very thin sheets before solidifying.

The seismic data which show that dust or fractured rock constitute much of the outer 10 or 20 km of the Moon apply also to the maria. However, the scarcity of craters on the maria makes it difficult to understand how lava could have been extensively fractured leave alone reduced to dust. It may be possible to reconcile the seismic evidence with fractured rock *beneath* the lava, which is no more than a few kilometres thick, particularly if the first few sheets that flowed were heavily fractured and were then covered by the sheets

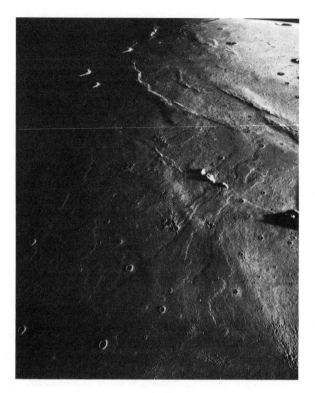

FIGURE 6.8
Part of the lunar Mare Imbrium, imaged obliquely by the Apollo 15 astronauts.

we see today. There is no doubt that the outer few tens of metres of the maria are fine dust, but this could be the result of many impacts too small to leave visible craters and of dust from large impacts elsewhere.

The scarcity of infill in the basins on the far side can be explained in the lava model *if* the Moon has a global crust and *if* it is thicker on the far side: the thicker the crust the less available would lava have been from beneath it.

There is no evidence that Plate Tectonics ever developed on the Moon, and therefore even if the maria are filled by lava the Moon has been far less geologically active than the Earth. This low level of activity indicates a lack of extensive interior melting, or a thick lithosphere, or a lack of stress in the lithosphere, or some combination of all three, all of which can result from a long history of low internal temperatures. As outlined earlier, low temperatures past and present are also indicated by the evidence that the Moon has a long history of outer rigidity.

There are no surface features on the Moon which suggest that the Moon ever had an atmosphere or oceans.

Topographic features are degraded by the impacts of bodies smaller than a few millimetres, by finely divided ejecta, by the solar wind and by seismic waves. It is also possible that surface dust is mobilized by electrical forces: indeed, such forces are of great importance to the minority of scientists who believe that very little lava if any has flowed from the lunar interior. I shall return to this in section 6.4. It has been estimated that these various small-scale erosional processes require tens of Ma to overturn a few millimetres of dust.

It seems very likely that apart form Moonquakes the Moon is geologically moribund today. However, ever since the invention of the telescope the occasional reddish glow and small patch of mist has been reported. These are called *lunar transient events* and their very transience makes it hard to determine whether all of them are illusory. Therefore, a very low level of geological activity today cannot quite be ruled out.

6.3.4 Lunar samples

About 400 kg of lunar material has been returned to the Earth from nine different sites on the Moon. Nearly all of this material was obtained by scooping the top of the surface, or by picking up small rocks up to about fist-size, or from cores a few centimetres in diameter and up to 2.4 metres deep. In only one case was a sample obtained from a large lump of rock, these being scarce on the lunar surface. This was at the Apollo 17 site where a sample was chipped from a boulder several metres across. It is a brecchia. Nearly all of the 400 kg consists of dust-size particles of rock. This dust is sometimes called *fines*, and it is also called soil but this is a poor term because it implies an organic component, which is absent. When it lies on the Moon the dust plus any small rocks is called the *regolith*. Brecchias are very common among the rock samples. Silicates are the dominant compounds.

The core samples, which consist of dust plus small rocks, exhibit layering. However, at a given landing site the sequence of layers in one core does *not* correspond to that in a neighbouring core, and therefore the subsurface material does *not* consist of a number of extensive thin sheets.

Many detailed studies have been made of the lunar samples. Here, in broad outline, are some of the more general and important conclusions.

At no site can the regolith be the broken-up product of a single type of rock. If there were originally regional differences in surface composition then this has now been lost because of the widespread scattering of impact ejecta and because of any other processes that can transport material over large distances. However, this mixing has not made the surface composition the same everywhere: this is clear from the lower albedo of the maria and from differences between the samples at one site and those at another.

The differences between the sites lie not so much in the dust, which has a broadly similar composition at all sites, but in the rocks. The clearest difference is between rocks from mare sites and those from highland sites. A greater proportion of the mare rocks resemble basalts on the Earth and are called *mare basalts*. Basaltic silicates are comparatively rich in iron, titanium and magnesium, and it is the extra iron and titanium that is largely responsible for the lower albedo of the maria. By contrast the highland rocks contain more silicates rich in calcium and aluminium.

The mare basalts are denser than the rocks which typify the highlands. But whereas the mare basalts are *too* dense to represent the lunar mantle, the highland rocks are not dense enough. This suggests that the mare basalts and the highland rocks have both been derived from mantle rocks of intermediate density. This possibility is supported by the lack of any known chemical pathway by which the mare basalts can be derived from the highland rocks. Mantle compositions that could yield both mare basalts and highland rocks are consistent with the constraints on the interior outlined earlier. Of course, this does not mean that the surface rocks *were* derived from the mantle.

The lunar samples show some striking differenes from Earth rocks. There is a broad tendency for the lunar samples, compared to the Earth, to be enriched in refractory materials and depleted in volatile materials. The depletion of volatiles is extremely marked in the cases of H, C, N and their volatile compounds as noted in Chapter 5. Moreover, the traces that *are* found can readily be accounted for by relatively recent contamination by the solar wind and by volatile-rich materials. There is evidence that volatile materials have *always* been scarce on the Moon. For example, there are no clays, no carbonates, no carbon compounds, and the presence of tiny particles of metallic iron in the lunar samples shows that the amount of free oxygen, such as would arise from the photodissociation of CO_2, has always been negligible. This scarcity of volatiles is in accord with the lack of any topographic evidence for a lunar atmosphere or oceans.

Among the refractory materials found in the lunar samples are refractory compounds of uranium and thorium. These elements are major sources of radioactive heating. If the surface abundances were typical of the whole Moon then, except for a very thin surface layer, the Moon would be entirely molten. Clearly the abundances of these materials at the surface exceeds those of the Moon as a whole and this could *either* be because of upward differentiation *or* because such materials were added late in the formation of the Moon.

6.3.5 *Dates of lunar surface events*

Most of the lunar highlands are close to being saturated with impact craters, notably on most of the far side and around the south pole (Figure 6.1). The remaining highlands bear fewer craters and this could largely be the result of partial burial by ejecta from basin-forming impacts, because these regions border such basins or the subsequent maria. Moreover, where these regions have been sampled brecchias are very common.

The number densities of craters on the maria are far lower than on the highlands, indicating, as outlined in section 3.3.4, that the maria surfaces are younger. Younger still are a few areas smaller than maria which bear yet lower number densities of craters.

Individual craters can be time-ordered, as outlined in section 3.3.4, on the basis of their state of degradation. In the case of the Moon this is aided by the existence of crater rays (section 6.3.1), as follows. Lunar surface materials are darkened by the combined action of the solar wind, by the bombardment of submillimetre particles, and by *cosmic-rays* which are high-speed fragments of atoms and which thinly pervade interplanetary space. Bright rays have not yet been darkened by these means and thus represent ejecta from craters of relatively recent origin. At the Apollo 16 landing site ejecta from the crater South Ray was examined for the effects of exposure and it is estimated to have been exposed on the surface for about 2 Ma. A crater called Bruno (after the Italian philosopher Giordano Bruno (1548–1600)) is particularly crisp, and it also has bright rays emanating from it. No close scrutiny and therefore no exposure age has been established. However, it is possible that a bright flash observed on the Moon by five English monks in the summer of 1178 was the formation of Bruno, in which case it is about the only lunar crater that has ever been observed to form.

Extensive radiometric dating of the lunar samples has been performed. The data are extremely complex and are not fully understood, but here are some broad conclusions.

Most of the highland-type rocks, even those found on the maria, have ages between about 3800 Ma and 4300 Ma, which suggests that a lunar-wide heating took place a little over 4300 Ma ago resetting the radiometric "clocks". Some highland-type rocks are a bit younger, down to about 3600 Ma, whilst others are a bit older. The oldest rock of all is 4600 Ma old and is a silicate type called *dunite*. This is not a typical highland rock but could be typical of the mantle.

Several of the mare impacts have been dated from the radiometric ages of the welding of the surrounding brecchias to which these impacts probably gave rise. The results are shown in Table 6.1. Also shown are the ages of the corresponding mare basalts, and you can see that they are several 100 Ma younger than the brecchias. *If* the samples of mare basalts are chips off basaltic lava infill then this infill occurred or was continuing several 100 Ma after the basin-forming impact.

Individual grains of dust are too small to be individually dated and thus each age determination of the dust samples corresponds to the average of many grains. The ages range from 4300 Ma to 4900 Ma, and dust from maria

The Solar System

TABLE 6.1
Ages of mare impacts (from brecchias) and mare basalts

Mare	Associated brecchia ages/1000 Ma	Mare basalt ages/1000 Ma
Serenitatis	4.3	3.8
Nectaris	4.3	—
Fecunditatis	4.2	3.5
Tranquillitatis	4.2	3.7
Humorum	4.2	—
Crisium	4.1	—
Imbrium	3.9	3.3
Orientale	3.8	—

sites displays much the same range as dust from highland sites. Most of the rocks are *younger* than this. Therefore, if the dust ages are correct then little of it could have come from the sort of rocks found in the lunar samples. However, it is possible that the dust ages are erroneously high because of the loss of the initial isotopes other than by radioactive decay (section 2.1.15).

But if most of the dust does indeed predate the rocks then its fairly uniform composition and age across the Moon could be because it was originally distributed uniformly from space. Otherwise, impact scattering and condensation from impact vapour must be responsible for this uniformity.

You can see that with the possible exception of a few dust samples there are no radiometric ages older than 4600 Ma, which is additional evidence for this being the age of formation of the planets.

No radiometric ages less than about 3000 Ma have been obtained for any lunar samples except for a tentative date of 850 Ma for material thought to have been ejected and heated by the formation of Copernicus (Figure 6.7). Thus, most of the lunar surface is extremely old: probably only about 1% of the surface is *younger* than 3000 Ma, whereas *less* than 1% of the Earth's surface is *older* than this. Much of the resetting of the lunar radiometric clocks is probably the result of impact heating, except that if the mare basalt samples are chips off lava infill then the resetting here would correspond to the solidification of geologically produced lavas.

From the ages of the samples it is possible to establish the length of time for which certain regions on the Moon, particularly the maria, have been exposed to bombardment. From the number density of craters on these regions it is thus possible to establish the lunar cratering rate over a considerable period of time as shown in Figure 6.9. Note the *logarithmic* scale of crater-production rate.

The solid line on the graph in Figure 6.9 extends from 3900 Ma ago to 850 Ma ago, this being the age range of dated regions on the Moon. It has been estimated that the cratering rate today (see section 8.1) is about 3 times what

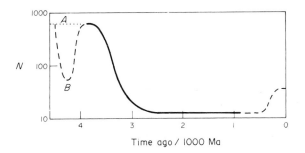

FIGURE 6.9

Impact crater production rates on the Moon, where N is the number of craters greater than 1 km diameter produced per 10^6 square kilometres per 10 Ma.

it was 850 Ma ago, and Figure 6.9 is brought up to today on this basis. Beyond 3900 Ma ago two possibilities are shown, labelled A and B. Curve A corresponds to the tail end of accretion of the Moon. The earliest part of curve B corresponds to this, then there is a lull, and then a broad peak in bombardment from a fresh supply of small bodies the possible nature of which I shall discuss in section 8.2.

In both cases it can be shown that the saturated areas of the Moon have ages of about 4200 Ma. It follows that few clues can be expected from the Moon about the development of its surface before about 4000 Ma ago.

From about 3900 Ma ago to about 3300 Ma ago it is clear from Figure 6.9 that the Moon was subjected to its last heavy bombardment. This is called the *late heavy bombardment*. During this period all the dated maria impacts occurred.

Between about 3300 Ma ago and about 2500 Ma ago you can see that the heavy bombardment gave way to a fairly constant light bombardment which persisted until comparatively recent times. This was probably the result of a separate population of small bodies from that which produced the earlier far heavier bombardments. Whether the increase in comparatively recent times points to a third population is a subject for Chapter 8.

6.4 The evolution of the Moon

6.4.1 Orthodox theories of lunar evolution

These theories are orthodox in the sense that they are believed by most lunar scientists. Here are the broad features of such theories.

About 4600 Ma the Moon accreted homogeneously from a medium that, relative to the Earth, was depleted in iron and in volatiles. If there were any volatile-rich bodies that made late impacts on the Moon then the weak gravitational field of the Moon would have allowed nearly all these volatiles to escape to space. Any core could have accreted first, or it could have later separated.

During the later stages of accretion, at the very least, the outer 100 km or so of the Moon melted, large areas being molten at the same time. From such melts the lower density silicates separated upwards to form a crust. The crust solidified and was heavily bombarded by later impacts, the bombardment tailing off about 3300 Ma ago. This bombardment fractured the crust.

To some extent the interior of the Moon cooled and the lithosphere became rather thick. But then interior temperatures began to rise as a result of radioactive heating. Some (or all) of the materials of the mantle melted and as the temperatures rose the upper boundary of the partially melted material also rose, and by about 3800 Ma ago was sufficiently shallow to provide surface lava through the mare basins which were weak points in the crust. The far side crust is thicker and therefore the basins gave rise to less weakening there, and consequently far less lava emerged into those basins than into those on the near side. By this time the interior was differentiated, and much of the radioactive material had moved to the surface regions (section 6.3.4).

When the abundances of the radioactive isotopes had declined considerably the interior began to cool again, the upper boundary of partially melted material sank to greater depths, and by about 300 Ma ago lava was no longer available to the surface. The interior was never sufficiently plastic close to the surface for long enough to cause widespread geological activity such as Plate Tectonics, largely because the Moon has a large ratio of surface area to mass and thus loses heat relatively rapidly (section 2.1.16).

The small rocks and fine dust have been produced by impacts throughout lunar history, and dust production has been enhanced by microbombardment. *Non*-brecchia rocks have been produced by impact-disruption of large sheets of various highland rocks and of mare basalts, and brecchias have been produced from fragments of these rocks.

You can see that much of what is known about the Moon can be accounted for by these theories. Indeed, they can account for the variety and juxtaposition of rocks in a rather more detailed way than the above outline suggests. However, these theories have to face certain major difficulties. Here are some of those that have been raised by various scientists.

1. The global scale non-hydrostatic state of the Moon cannot be the result of a largely molten outer Moon solidifying in any gravitational environment to which the Moon is likely to have been exposed.

2. Some astronomers think that the more rapidly a body accretes the hotter it gets, and that for the outer Moon to have melted it is necessary for the Moon to have accreted in roughly 0.01 Ma. Therefore, orthodox theories cling to curve *B* in Figure 6.9, because curve *A* is associated with protracted accretion. However, theoretical and experimental studies of the accretional

processes show that it is unlikely that a body with the low mass of the Moon could have accreted in less than about 10 Ma. This would not result in extensive melting of the outer 100 km or so. The late heavy bombardment, whether it was the end of accretion or a separate event, is even less likely to cause widespread melting. Radioactive heating can help overcome the likely ineffectiveness of accretion in melting the Moon, particularly if there were short-lived quick-acting isotopes in the early Solar System such as ^{26}Al (Al = aluminium). However, even with their aid it is thought by some to be unlikely that the outer Moon could have been melted quickly enough to yield those highland rock samples that have given radiometric ages of about 4300 Ma and older.

3. It is hard to see how global melting of the outer Moon could yield a crust with a different thickness on one side of the Moon to that on the other.

4. Most of the mare basins were in place by the time Mare Serenitatis was filled with lava (Table 6.1). The lava was therefore available and yet it did not fill, for example, Mare Imbrium until 500 Ma later.

5. Layered bodies of rock, such as might have come from a succession of lava flows, are rare on the Moon, and those observed might be the result of other processes, such as impacts.

6. The radiometric ages of the dust exceed those of the rocks from which the dust is supposedly derived. There are plausible reasons for adjusting the dust ages downwards, but some scientists find these reasons uncompelling.

All theories have difficulties to overcome, and there is certainly no need to abandon the orthodox view. But there is an alternative theory of lunar development which overcomes the difficulties outlined above, and which also overcomes several other difficulties, though it does have its own particular set of difficulties to face.

6.4.2 An unorthodox theory of lunar evolution

This theory is largely due to Thomas Gold (b. 1920), an Austrian-born astronomer who has spent most of his working life in Britain and in the USA. I have called it an unorthodox theory because few adhere to it. However, this does *not* mean that it is wrong.

In a nutshell, orthodox theories are based on homogeneous "hot" accretion, the high temperatures leading to differentiation. This unorthodox theory is based on inhomogeneous "cold" accretion, the only differentiation being the separation of any lunar core that might exist because of the increase in temperature expected towards the centre of any planet (section 2.1.16). Any boundary at a depth of about 300 km (Figure 6.3) would be the direct result of inhomogeneous accretion. The "crust-mantle boundary" at a depth of 50 km on the nearside could be confined to mare basins, the result of the mare impacts. If it is a global feature then it too could be the direct result of inhomogeneous accretion, or of the final stages of compaction of accreted bodies under an overburden that grows with depth. The same two kinds of explanations could also apply to the layering of the outer few kilometres of the Apollo 17 site.

The 5 km offset of the centre of figure from the centre of mass need not be explained by a change in crust thickness from 50 km on the near side to 75 km on the far side, but by an asymmetry spread over a greater depth and consequently slight on a global scale.

The surface of the Moon is the result of accretion and of subsequent impacts and slow-acting erosional processes. There are no flows of lava from within, the only liquid rock at the surface being the local result of impacts. Compounds bearing radioactive elements are acquired late, and are therefore confined near to the surface thus avoiding the problem that would otherwise be posed by the observed surface abundances (section 6.3.4).

Small rocks would survive impact, with some fragmentation. Larger rocks would yield molten material and brecchias. Clearly a great variety of rock types would result. The variety at the surface is greatly increased by the excavation through impacts of different layers of accreted material, thus converting an initial vertical variation in composition into the horizontal variability observed. The greater abundance of mare basalts on the mare than in the highlands is *not* explained in any detail, but the greater variability of rocks *within* the highlands could be the outcome of the maria impacts which excavated the Moon to large depths and thus brought up a great variety of layers.

Dust is expected to be common in the PFM, and in this theory the direct accretion of such dust accounts for its abundances at the lunar surface, though smaller quantities are tiny impact-fragments and the result of condensation from rock vapour generated by impacts.

The observed uniformity of dust on the near side of the Moon is to be expected if most of this dust fell from space. Moreover, the generation of dust by impacts would not disturb this uniformity. You might anticipate that dust *from space* accounts for the mare infills, but this is not the case. By the time the mare impacts occurred there would probably have been too little dust available to fill the maria. Moreover, infall alone would not level the mare floors. It is therefore proposed that the mare infills are the result of dust migrating into the mare basins from elsewhere on the Moon. Such migration need not disturb the uniformity of the dust type across the lunar surface. Clearly, the dust must have been far less mobile on the lunar farside.

Theoretical studies and laboratory experiments have indicated that lunar dust could be mobilized by electrical forces which on the Moon could arise from electrically charged particles which impinge on the lunar surface. The difference in dust mobility between the two sides of the Moon could be the result of the Earth's magnetic field as modified by the solar wind, which, as outlined in section 9.1.1 produces a magnetotail stretching away from the Earth, roughly in a direction pointing away from the Sun. This magnetotail could steer larger quantities of charged particles on to the near side than on to the far side.

This mobilization continues today and can explain many features on the Moon. Try for yourself, in relation to Figure 6.8, the "Gestalt switch" from lava scarps to a very slow-moving surf of dust.

When the mare basins were excavated the fragmentary material beneath them must have been compressed to a higher density than before. Thus, even if, as seem *un*likely for a cold Moon, isostatic equilibrium was subsequently achieved there would still be a residual basin. The dust infill then produces the mascon, borne by the cool rigid outer Moon.

The unorthodox theory also provides a satisfactory explanation of exposure age. When a cosmic-ray encounters solid matter then in most cases it penetrates only a few millimetres and leaves a detectable track. In the case of a dust grain the number of tracks indicates for how long the grain has been *un*shielded by other material. This is the *exposure age*. Throughout the lunar core samples, which extend well below a few millimetres, the exposure ages of the grains are similar at all depths and correspond to a few million years exposure to the cosmic-ray intensity in the *present* Solar System. In this unorthodox theory this is interpreted as the typical time for which a dust grain moved through space after it was formed and before it accreted to the Moon. Orthodox theories argue that the turn over of the dust at the lunar surface has produced the observed uniformity of track densities. But this turnover is so extraordinarily slow that each grain could only have been exposed during the last 4600 Ma for a total time much less than a million years thus requiring considerably higher cosmic-ray intensities in the past, which though possible is thought by several scientists to be unlikely.

The greater radiometric age of the dust is readily explained by the unorthodox-theory, and so too are the great ages of the rocks. However, it is not clear whether the ages of the mare basalts date the infill by *dust*. If not then though the general features of the graph in Figure 6.9 remain, they are based on fewer data.

The greatest difficulties faced by this unorthodox theory are to do with the compositional variations across the Moon. The greater abundance of mare basalts on the maria, their smaller ages than most highland rocks, the possibility of deriving the highland type rocks and the mare basalts from a common source material in the mantle, and many other compositional relationships, are glossed over in an unsatisfactory manner. It remains to be demonstrated that this theory can overcome such difficulties.

6.5 The origin of the Moon

The aspect of lunar origin with which I shall be primarily concerned here is how the Earth came to get its Moon.

There are three types of theory of how the Earth came to get its Moon: fission theories; capture theories; and binary accretion theories. But first, some basic science.

● 6.5.1 Angular momentum

The angular momentum of a system is a measure of its rotation. A point has to be specified with respect to which the rotation is measured. In Figure 6.10 this point is P, and in our case would typically be the centre of a planet. Suppose that a small mass m is made to go around P in a circular path radius r as shown and that its speed is s. The *angular momentum* of m is *defined* to be the product of m, r and s. Any body can be thought of as consisting of lots of small masses m, and the angular momentum of such a body around a point P is obtained by combining the products mrs according to certain rules which I shall not describe.

FIGURE 6.10
Illustrating angular momentum.

A very important property of angular momentum is that if no angular momentum is transferred to or from a body then, regardless of any *internal* changes in the body, its angular momentum remains fixed. This is the *principle of conservation of angular momentum*. For example, if r is somehow reduced in Figure 6.10 without transferring angular momentum to or from m then, because the angular momentum mrs is constant, s must increase.

6.5.2 Fission theories

Fission theories rely on the rotational disruption of the Earth. There are several types of fission theory. The type I shall outline is illustrated in Figure 6.11.

In Figure 6.11 (a) the Earth is rotating *almost* fast enough to begin to lose mass from its equatorial regions. The iron core of the Earth has partially segregated, and therefore the outer regions are depleted in iron. Further gravitational segregation of the iron core means that denser material moves inwards displacing less dense material outwards. The effect on the Earth's rotation is the same as moving the mass m closer to P in Figure 6.10: the Earth rotates faster. The 1% of the Earth's mass necessary to form the Moon

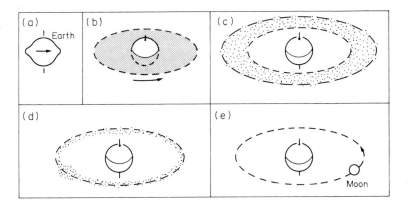

FIGURE 6.11
A fission origin of the Moon. The eccentricity of the material around the Earth is
the result of the oblique viewpoint.

is thus flung off. It is assumed that the Earth is hot enough for its outer regions to have previously been vaporized, in which case it can be shown that the material flung off forms a thin disc of gas as in Figure 6.11 (b). The gas cools, materials condense and accrete to form the Moon as illustrated in Figure 6.11 (c), (d) and (e). Because of core segregation in the Earth the Moon is depleted in iron, and because of the vaporization of the outer mantle of the Earth a degree of chemical segregation is plausible by means of which the Moon comes to consist largely of materials that differ to the required extent from those of the Earth's mantle.

There are, however, two grave difficulties facing all fission theories.

First, all fission theories place the Moon in an orbit of very low inclination with respect to the Earth's equator. This inclination today is 5.15°, and no convincing explanation has yet been found of how the inclination could have increased to this value from about 0° 4600 Ma ago.

Second, the axial spin rate of the Earth 4600 Ma ago can be estimated using the principle of conservation of angular momentum by giving the Earth all the present angular momentum of the Moon plus the orbital angular momentum of the Earth and the Moon around their centre of mass: the Earth spins *nowhere near* fast enough to have spun off lunar material and no very plausible means have been suggested whereby the Earth–Moon system could since have lost the necessarily copious amount of angular momentum.

6.5.3 Binary-accretion theories

In these theories the Earth and the Moon accrete separately from a *common* pool of dispersed materials. For various reasons the accretion of the Earth proceeds more rapidly, leading to a situation resembling that in Figure 6.11 (c) but without this material having to have an inclination close to 0°. Moreover, the Earth need never have been spinning particularly rapidly. Thus, the two grave difficulties facing fission theories do not arise.

The Moon accretes from the ring of material.

These theories can account in broad terms for the compositional differences between the Earth and the Moon, but only by making a number of fairly detailed assumptions about the conditions in the PFM in the region where the Earth and the Moon formed.

6.5.4 Capture theories

These theories yield the compositional differences between the Earth and the Moon by forming their materials in widely separated parts of the PFM. This requires few assumptions about conditions in the PFM. The Earth then captures lunar material.

In the most dramatic of these theories the Earth captures the Moon after the complete Moon has accreted elsewhere. Capture is most readily achieved if the Moon and the Earth approach each other at fairly low speeds. Unfortunately this means that they occupy fairly similar orbits and could not have been formed in widely separated parts of the PFM.

At higher speeds it is difficult to dissipate sufficient of the energy of motion of the Moon for it to be captured and at the same time yield an Earth–Moon system that resembles the one that actually exists.

Capture is made considerably easier if the Moon was captured in small pieces, because interactions between them, or between circum-Earth gases and such pieces would readily place some of them in orbit around the Earth. This again leads to a situation resembling Figure 6.11 (c) from which the Moon accretes. However, the greater the ease of capture the more difficult it becomes to explain why the Earth has a large satellite but not Venus, Mercury and Mars.

In one version of such theories the pieces are of submillimetre size. They form well beyond the Earth and spiral inwards through the *Poynting-Robertson effect*. You can think of solar radiation as consisting of a rain of tiny bullets moving outwards from the Sun at the speed of light. For a body moving in orbit around the Sun the surface facing the direction of orbital motion will be struck by more "bullets" than the trailing face, just like the front of you gets wetter than the back when you run through a storm of vertically descending rain. The body in orbit is thus slowed down slightly. This is the Poynting-Robertson effect, and is named after the British physicist John Henry Poynting (1852–1914) and the American physicist Howard Percy Robertson (1903–1961). Any reduction in orbital speed by the action of an additional force can be shown to lead to shrinkage of the orbit. Thus, the body spirals inwards. In the Solar System the Poynting-Robertson effect will produce appreciable inward spiralling only of bodies of submillimetre size.

Recent evidence based on oxygen isotope ratios in the Earth and in lunar samples suggests that the Earth and the Moon may *not* have formed in widely separated parts of the PFM. However, the extent to which this poses a difficulty for capture theories is not clear.

6.5.5 *The late heavy bombardment of the Moon*

In many fission theories the late heavy bombardment is a consequence of the final stages of accretion of the Moon from material within the Earth–Moon system. It is observed that the maria and the large unfilled basins tend to lie in the equatorial regions of the Moon. This could be the result of a tendency for the larger bodies in the final stages of accretion to lie in the orbital plane of the Moon, with the Moon already possessing an axial inclination similar to that of today. Moreover, *if* the Moon was already in synchronous rotation around the Earth, and *if* there was more spare material *between* the Earth and the Moon than *beyond* the Moon, then the rather greater number of large basins on the near side than on the far side can be explained. The several hundred Ma delay in the basin impacts after the main phase of accretion can also be explained, because it can be shown that the more massive a body the longer it is likely to survive before being collected by the Moon (or by the Earth).

In some binary accretion theories the story is much the same. In others of these theories the large basins are produced by bodies which come from *beyond* the Earth–Moon system, in which case the smaller number of large basins on what is now the lunar far side has to be put down to chance, which is just plausible. In these theories the late heavy bombardment can be separate from accretion, as in curve *B* in Figure 6.9.

In capture theories more possibilities open up: the Moon could have suffered its late heavy bombardment as it tumbled through space before capture; it could have suffered it from debris already in orbit around the Earth or the large basins alone could have arisen from such debris; or it could have been bombarded by material which entered the Earth–Moon system *after* the Earth had captured the Moon.

6.6 Summary

Though the Earth and the Moon can be regarded as a double planet system this is *not* an association of twins. The Moon contains less iron than the Earth, far less volatiles, and, at least on its surface, a greater abundance of refractory materials. There seem also to be other compositional differences, such as in the silicates which make up the greater part of both planets.

The Moon has been far less geologically active than the Earth, and there has been no weathering. Consequently, the surface is peppered with impact craters, many of which date back to within a few hundred Ma of the Moon's formation. Homogeneous "hot" accretion of the Moon with subsequent differentiation and subsequent cooling is currently the more fashionable view of lunar development than heterogeneous cold accretion. *If* the Moon were ever hot then it would tend to cool rapidly because of its small size and consequent large surface area to mass ratio.

There are many viable theories of how the Earth came to acquire its Moon, reflecting the considerable uncertainty as to how it actually happened. It is correspondingly uncertain as to why its composition differs from that of the Earth.

6.7 Questions

1. In what *one* way does a *tidal* force differ from any other type of force?
2. Suppose that the Moon has a crust which solidified when an appreciable tide, due to the Earth, existed in it. Why is this *not* an explanation of the lunar crust being thicker on one side than on the other?
3. An artificial satellite is in the Keplerian orbit of co-rotation above the Earth's equator. What is its sidereal orbital period around the Earth?
4. Two rocky bodies, masses m_1 and m_2, densities d_1 and d_2, are in orbit around a planet of mass M. Both bodies are several hundred kilometres diameter. If m_1 is *greater* than m_2 and d_1 is *less* than d_2, which of m_1 and m_2 has the closest Roche limit to the planet? Why is the diameter of the bodies relevant?
5. List the items of evidence which indicate that the lunar interior is cooler today than that of the Earth, and probably always has been.
6. List the main differences in chemical composition between the Earth and the Moon.
7. Name *seven* different forms of ejecta from impact craters, and *four* topographic features that impact craters can bear. What factors influence the type of crater and associated features formed by an impact?
8. Outline how the lunar cratering rate over much of the last 4600 Ma has been estimated.
9. Prepare a *short* table showing the features of the Moon readily explained by orthodox theories of lunar development and those readily explained by unorthodox theories. List also the main difficulties faced by each type of theory.

10. If, without any transfer of angular momentum to the Earth–Moon system, the Moon were shifted into a circular orbit half the radius of its present orbit (which is nearly circular), discuss whether its sidereal orbital period would *initially* increase or decrease.

11. In relation to each of the three types of theory of how the Earth got its Moon, outline how the late heavy bombardment might have happened.

7

MERCURY

The orbit of Mercury, like that of Venus, lies within the orbit of the Earth. Indeed, it is the closest planet to the Sun and at its maximum elongations is never more than about 28° from the Sun. This makes Mercury hard to see from the Earth. "Mercury" is the name of the Roman messenger of the gods: an elusive figure.

By the middle of the nineteenth century it had been firmly established that there was something curious about Mercury's orbit. With respect to the distant stars the major axis of the orbit moves in the prograde direction at the rate of 574 seconds of arc (") per century, as shown for 200 centuries in Figure 7.1. This is called *precession of the perihelion*, and is not in itself curious because it had long been known that precession of the perihelion is one result of the gravitational influence on a planet's orbit of all the other planets in the Solar System. But when allowance was made for the influences of all the other planets on Mercury then there was a residual precession unaccounted for of 43" per century, as shown for 200 centuries in Figure 7.1.

One of the earlier attempts to account for this residual precession was by the astronomer Le Verrier, to whom I shall introduce you more properly in Chapter 13. In 1859, from the residual precession he predicted that within the orbit of Mercury there either existed several small bodies, or a single planet. Soon afterwards it was believed that such a planet had been discovered in transit across the Sun. It was named Vulcan. However, its

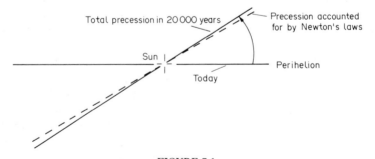

FIGURE 7.1
Precession of the perihelion of Mercury. The viewpoint is perpendicular to the orbital plane.

existence was never firmly established, and today it is believed not to exist.

In 1915 the German physicist Albert Einstein (1879–1955) published his *General Theory of Relativity*. This had grown out of his very successful *Special Theory of Relativity* which he had published in 1905 and which can account for two-thirds of the 43″ residual. The Special Theory replaces Newton's Laws of Motion and the General Theory also replaces Newton's Law of Gravity. With the advent of the General Theory the whole of the 43″ residual could be explained to within observational uncertainties, and this was a major success for the General Theory, a success it has since built on. Nevertheless, since 1915 other theories of gravity have been advanced, though the great majority of scientists today accept Einstein's General Theory and the explanation that it offers of Mercury's precession.

The residual precession is greater the closer the planet to the Sun and the larger its orbital eccentricity. For these reasons the residual precession is greater for Mercury than for any other planet. It is insignificant for all satellites because the residual precession is smaller the smaller the mass of the primary body, and no planet is anywhere near as massive as the Sun. For nearly all purposes in the Solar System the far simpler Newton's Laws are adequate.

7.1 The exploration of Mercury

Before the advent of radar the best measurements of Mercury's radius were made when Mercury was in transit across the Sun. The result was 2440 km. The mass of Mercury was *first* determined from its influence on the orbit of Eros, an asteroid which was discovered in 1898 and which often passes close to Mercury. The calculated mass was 0.056 M_E, which gave a mean density of 5500 kg/m^3.

Early in the nineteenth century it was wrongly deduced, on the basis of scanty observations of surface features, that the axial sidereal period was about 24 hours, prograde. Mercury is a small planet, which combined with its small angular separation from the Sun makes it a difficult planet to observe. Therefore, it was not until 1890 that this erroneous period was revised, not by an hour or so, but to about 88 days prograde. This was sufficiently close to the sidereal orbital period of 87.97 days for it to be generally concluded that Mercury was in synchronous rotation around the Sun, and therefore that the sidereal axial period was also 87.97 days.

The earlier period of about 24 hours seems to have resulted from the observation that the elusive surface markings of Mercury lay in roughly the same positions on consecutive nights. Clearly this is consistent both with a period of about a day and with a very long period.

Then, in 1965 radar echoes from Mercury showed that the planet did not after all have a period of 88 days, but about 60 days. Subsequent radar observations, plus Earth-based telescope observations guided by the 60-day period, ultimately established beyond doubt that the axial sidereal period of Mercury is 58.64 mean solar days. The spacecraft Mariner 10 flew by Mercury in 1974 and 1975. It confirmed this period, but did not improve on its accuracy.

It can be shown that many of the observations adduced to support the 87.97-day period also support the 58.64-day period, largely because these periods have the simple ratio of 3/2. Nevertheless, there were older observations that would long ago have shown that the longer period was wrong. Doubtless these were given too little weight because it was reasonable to expect that tidal forces of the Sun had given Mercury synchronous rotation.

Indeed, Mercury *is* in a form of synchronous rotation around the Sun. The 3/2 ratio of Mercury's orbital and axial sidereal periods means that, as illustrated in Figure 7.2, for every orbit of the Sun Mercury rotates one and a

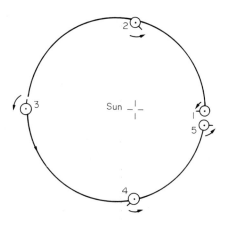

FIGURE 7.2
The axial rotation of Mercury. The line on Mercury is fixed to its surface at the equator.

half times around its axis. You can see that it therefore presents the same face to the Sun at *every other* perihelion and that any tidal bulge would lie along the Sun–Mercury line *every* perihelion. Mercury's orbit is fairly eccentric, and therefore a tidal bulge would be acted on by the Sun a good deal more strongly near perihelion than elsewhere in the orbit. The tidal braking would therefore be greatest around perihelion, and it is this that has presumably led to the partial synchronization observed.

Radar observations have also shown that the inclination of the spin axis is less than 3°. This is consistent with Earth-based telescope observations. Observations by Mariner 10 have not enabled this upper limit to be reduced but if the ³⁄₂ synchronization is stable then the axial inclination must be much closer to zero.

Mercurian surface temperatures have been determined from observations of thermal radiation from the planet's surface. The hottest place is at the equator at perihelion, which, because of the ³⁄₂ synchronization, at all perihelia is one of two places on the surface. The temperature is about 700 K, about the same as the surface temperature everywhere on Venus. On the night hemisphere much of the surface is at about 100 K. Were Mercury to keep the same face to the Sun then the night hemisphere would be close to 0 K. The higher temperature that actually exists has been known since 1962. At that time it was still believed that Mercury did keep the same face to the Sun, and therefore this temperature was used to support the idea that Mercury had an atmosphere, because such an atmosphere would transport heat around from the Sun-facing side, thus raising the night-side temperature. There was, however, little other evidence that Mercury had an atmosphere, and therefore when the correct axial rotation period was established in 1965 most astronomers quickly settled on the correct view that Mercury has no appreciable atmosphere.

Radar observations yielded a radius for Mercury of 2439 km which has an observational uncertainty about 50 times less than earlier values. Radar also reduced the uncertainty in the mass, by a method similar to that used for Venus (see section 4.1.2).

In 1974, just before the spacecraft Mariner 10 flew by Mercury, the best Earth-based optical resolution of the surface was about 300 km. This means, for example, that a feature on Mercury the size of Sri Lanka would be a featureless blur. Such indistinct features were, in the end, good enough to accurately establish the axial rotation rate but tell us little about the nature of the Mercurian surface.

Radar was unable to improve on the optical resolution of surface features. However, radar reflection spectra did indicate that the surface of Mercury consists largely of dust to a depth of at least a few metres, and that the dust could be composed largely of silicates. These conclusions are supported by the *optical* reflection spectra, and by polarimetric and photometric analyses of the thermal radiation emitted by Mercury, day and night. These radar and optical data are rather similar to corresponding data for the Moon. Mercury is also similar to the Moon in having a Bond albedo of about 7%.

Mariner 10 is the only spacecraft to have yet visited Mercury. It was launched on 3 November 1973, passed fairly close to Venus on 5 February 1974 returning pictures and other data from that planet, and flew by Mercury on 29 March 1974. The orbit of Mariner 10 around the Sun is such that it passes close to Mercury about every 2 × 88 days. Data were returned from the first fly-by and from the next two, on 21 September 1974 and 16 March 1975. Then, the fuel required to adjust and orientate the spacecraft was used up, and it now tumbles uselessly around its orbit.

Because the Mariner 10 orbital period is about twice the orbital period of Mercury, and because of Mercury's own ¾ synchronization, much the same hemisphere of Mercury was sunlit during each of the three fly-bys. The optical images are thus of half a planet.

The images were returned by the on-board TV cameras and yielded detail on the Mercurian surface down to about 1 km across, about as good as the Moon viewed through a telescope with a magnification of $\times 120$ and about 300 times better than the pre-Mariner 10 images. Parts of the surface were imaged with a resolution of about 0.1 km.

The surface is dominated by impact craters. This came as no surprise: the lunar-like data outlined above, the small size of Mercury and the corresponding likelihood of little geological activity, the absence of a significant atmosphere, the abundance of craters on Mars, all led to the expectation that Mercury would have craters.

No satellites of Mercury were discovered, and therefore any satellite cannot exceed about 3 km radius.

7.2 The interior of Mercury

The occultation by Mercury of the radio signals from Mariner 10 yielded a radius of 2440 km, which is no improvement on the radar value. However, from the gravitational influence of Mercury on the orbit of Mariner 10 a mass was obtained with 5 times less observational uncertainty than radar had yielded, to give a mean density of 5430 kg/m³. This lies between the mean densities of Venus and the Earth, but because of the small mass of Mercury the uncompressed density is not much less than 5430 kg/m³, whereas the uncompressed densities of Venus and the Earth are considerably less than the planetary values. Mercury thus contains a much *greater* abundance of denser materials, the strong implication being that it is rich in metallic iron. If Mercury consists of a largely iron core and a largely silicate mantle, then the core would account for about 80% of the mass and about 40% of the volume, which would make Mercury richer in iron than any other planet. An alternative is that it consists of a few per cent silicates, all the rest being iron-rich compounds such as FeS and FeO, which have densities between those of metallic iron and silicates. This, however, makes silicates and iron surprisingly scarce and FeS and FeO surprisingly abundant.

The flattening f of Mercury is very slight. The several occultations by Mercury of the radio signals from Mariner 10 yielded radii of cross-sections inclined at about 2° and 68° with respect to the equatorial plane, and the radii are the same to within about 0.5 km. Only an upper limit on the flattening can thus be established, but this is a good deal less than the flattening of the Earth. This is not surprising, because of the slow axial spin of Mercury.

These measurements also make it clear that Mercury is a good deal more spherical than the Moon, perhaps because it has not been subjected to the tidal forces of a nearby planet.

Only upper limits have been established for J_2, the smallest being from Mariner 10. This upper limit is very small, and this is consistent with the slow axial spin and nearly spherical shape of Mercury. This nearly spherically symmetrical mass distribution helps explain why the synchronous rotation of Mercury is only partial (compare section 4.2). An upper limit for J_2 can also be obtained by calculation from the upper limit of f and the axial spin period T_a, on the assumption of hydrostatic equilibrium. The calculated value is *less* than the upper limit from Mariner 10, indicating that Mercury is close to hydrostatic equilibrium. However, if hydrostatic theory is then used to calculate the possible range of values of the polar moment of inertia C from the axial spin period T_a and the allowed values up to the upper limit of either f or J_2, the calculated range of values of C is so wide that no useful constraints on the variation of density with depth are obtained.

There are no data of any significance on smaller-scale gravity variations.

Observations by Mariner 10 have demonstrated that Mercury *does* have an intrinsic magnetic dipole moment. Its value is about 0.00005 of that of the Earth, it is inclined about 12° from the spin axis, and its polarity is the same as that of the present Earth's field. The existence of such a field came as a surprise. Recall that planetary dipole fields are thought to require liquid conductors plus rapid axial spin. When Mariner 10 reached Mercury the following data on planetary magnetic fields were to hand: the Earth rotates quickly, almost certainly has a large molten iron core, and has a strong dipole field; Mars rotates quickly, but probably has no molten iron core, and has no detectable dipole field; Venus probably has a sizeable molten iron core but rotates slowly, and has no detectable dipole field (perhaps also because it has no solid *inner* core — see section 4.2); the Moon rotates slowly, probably has no liquid iron core, and has no detectable dipole field. Therefore, though the high density of Mercury suggested it might well have an iron core, Mercury was known to spin slowly and its small size indicated that its interior was cool enough for most or all of any core to be solid, and thus no detectable dipole field was expected.

It has been suggested that the magnetic field observed outside Mercury and which is consistent with an interior dipole source, arises instead from magnetized rocks near the surface, which happen to produce a net field which mimics that of an internal dipole source with the observed and expected low inclination with respect to the spin axis. This could be the result of an earlier internal field, lost when Mercury cooled.

It had also been suggested that the field is a remnant of a time when Mercury had suffered less tidal slowing of its axial spin, at which time the core was molten, Mercury having cooled less than today. In this case a substantial internal magnetic dipole moment is expected. As the core cooled it would form a solid outer shell, and as this shell further cooled through the *curie temperature* it could retain today the magnetic field of the remnant liquid iron core within it. The Curie temperature is simply the temperature above which a substance cannot retain a magnetic field in the absence of a generating process, and is named after the French physicist Pierre Curie (1859–1906). For iron it is about 1000 K. This is a *plausible* explanation of the Mercurian field, though it is not certain that a slowly cooling core would end up with an appreciable magnetic dipole moment.

However, Mercury's magnetic dipole moment does indicate that Mercury has an iron core, and leaves open the possibility that some of the core may still be molten. It could even all be molten if Mercury possesses a powerful source of internal heating, and this would certainly remove some of the mystery surrounding the dipole field's origin.

Figure 7.3 is a model of the Mercurian interior consistent with the various

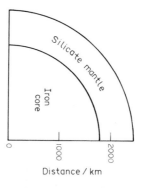

FIGURE 7.3
A model of Mercury.

(weak) observational constraints. The core and mantle could be the result of heterogeneous accretion, iron first then the silicates, or it could be the result of homogeneous accretion with the Mercurian interior becoming sufficiently hot for differentiation to subsequently occur. The latter mode, as you saw in relation to the Earth (section 2.1.16), has the possible difficulty that the substances which would ultimately form the mantle may lose oxygen to iron, but until the present oxygen abundance of Mercury's mantle is known it will not be known whether this possible difficulty applies.

If any Mercurian core separated from the mantle then it is likely that mantle temperatures became high enough for differentiation *within* the mantle to occur, the less dense substances moving upwards to form a crust. At present it is not known whether Mercury has any such crust.

7.3 The surface of Mercury

7.3.1 Impact craters

Figure 7.4 shows the two halves of the sunlit hemisphere imaged by Mariner 10. The preponderance of impact craters is plain. That nearly all are impact craters is clear by comparison with the impact craters on the Moon, with which they bear a striking similarity. A few small rimless bowls may be volcanic.

There are, however, some differences between cratering on Mercury and the Moon. For example, ejecta that can be associated with a specific primary impact lie much closer to the primary than would have been the case on the Moon. I refer, in particular, to ejecta blankets and to secondary craters. This is clearly a result of the higher surface gravity on Mercury, a little over double that on the Moon.

The older large craters can be identified by the relatively large number of small craters which they bear. They differ from corresponding lunar craters by having no central peaks and by having rather smooth rims. This is a kind of degradation which indicates that when they were formed the Mercurian surface may have been warm enough for it to be slightly plastic. However, none of the other observed differences between Mercurian and lunar craters can be used to argue that there are, or have been, any other fundamental differences in the composition and nature of the lunar and Mercurian surfaces. The surface of Mercury, like that of the Moon, could consist largely of fragmented silicates, even to the point of dust.

The surface of Mercury, broadly speaking, can be divided into two types of terrain, *heavily cratered terrain*, which accounts for about 80% of the imaged hemisphere, and *smooth plains*, which accounts for the remainder.

Heavily cratered terrain can be subdivided into two types. First, there are areas in which the number–density of craters approaches saturation. Second, and more commonly, there are less heavily cratered areas in which there are fairly well-spaced craters larger than about 20 to 30 km diameter, the spaces between being called *intercrater plains*. This is a rather unfortunate name, which apparently arises from early low resolution pictures which suggested that they were flat areas devoid of craters. By contrast these areas are rolling plains bearing many craters smaller than about 10 km diameter plus a few craters with diameters in the range 10 to 20 km. Most of the smaller craters on the intercrater plains are secondaries, as indicated by their shallow elongated forms and their tendency to form clusters and chains. The lower half of Figure 7.5 is a fairly typical example of this second kind of heavily cratered terrain.

Close scrutiny of this second kind of heavily cratered terrain reveals that the subsaturation level of cratering is not the result of the partial obliteration of a once saturated surface, but the partial reworking of an old and largely

(b)

(a)

FIGURE 7.4

Mercury imaged by Mariner 10, (a) the approach view, (b) the departure view. These are mosaics from pictures taken at ranges of a few thousand km. North is at the top.

FIGURE 7.5
Mercurian smooth plains (top) and intercrater plains (bottom) imaged obliquely
by Mariner 10. The width of the picture is about 490 km.

crater free surface. It is possible that this surface, which is not entirely obliterated in the intercrater plains, predates almost all the craters on Mercury. The surface must surely have suffered saturation bombardment during the final stages of accretion but either the surface was too soft to preserve these craters, as indicated by the degradation of the oldest craters, or the surface was somehow "decratered".

The Moon has no analogue of the Mercurian intercrater plains. The maria and small plains in the highlands can readily be seen to *postdate* the craters on their borders. *If* the Moon also lost most of its craters from the final stages of accretion then the apparent failure on Mercury of post-accretional bombardment to saturate the intercrater plains may seem to indicate that Mercury was exposed to a less numerous population of small bodies than the Moon. However, this need not be the case: the lower surface gravity of Mercury could be responsible because it confines ejecta blankets and secondary craters closer to the primary than is the case on the Moon. Of course, there is no direct evidence that the Moon *has* lost an appreciable fraction of its craters from the final stages of accretion, except through any separate late heavy bombardment which merely replaces one set of craters by another.

Smooth plains are distinguished from intercrater plains by a far lower number density of small craters and by clear evidence that on the smooth plains there has occurred the obliteration or partial obliteration of craters. Both of these features should be apparent in the upper half of Figure 7.5, which shows a smooth plain. Smooth plains material is younger than most of Mercury's craters, but is not as lightly cratered as the lunar maria.

Some smooth plains consist of infill in some medium sized craters. The largest known expanse of smooth plain is an infill of the largest multi-ringed basin (section 6.3.1) known on Mercury, called *Caloris*. This is just visible near the centre of the night–day boundary in Figure 7.4 (b), and more clearly in Figure 7.6. It is one of the few features that can be identified on Earth-based drawings of Mercury. The inner ring, *Caloris Montes*, is about 1300 km diameter and consists of lumpy terrain reaching to about 3 km above the adjacent plains. A second, less distinct ring is visible further out.

FIGURE 7.6
The Caloris basin on Mercury (middle left) imaged by Mariner 10. North is at the top.

Much of the area in and around Caloris is smooth plains infill. This infill bears few craters, and only a few craters have been partially obscured, indicating that the whole Caloris basin since its creation has suffered few impacts and is therefore comparatively recent. The plains are, however, scarred by ridges about 1.5 to 12 km wide and about 0.5 to 0.7 km high, and by cracks about 8 km wide and about 0.7 km deep (Figure 7.6). The cracks are predominantly concentric with respect to the impact centre, though some are radial. The ridges are more freely distributed in orientation. If the smooth plains infill contracted then cracks are to be expected. Other "cracks" are probably grooves cut by impact ejecta, and some of the ridges may be ejecta deposits. However, the origin of most of the ridges and cracks is a mystery. Some short scarps and hummocks in this region probably also originate from the Caloris impact.

Diametrically opposite from the impact centre of the Caloris basin lies an area of lumpy terrain, consisting of hillocks about 2 km tall and hollows of about the same depth. There are also a number of short, straight scarps. This topography is probably the result of seismic waves radiating from the Caloris impact, which would be refocused on the opposite side of Mercury. On the Moon similar disruption of the points opposite the Imbrium and Orientale impact centres is seen.

7.3.2 Other surface features

There is no very clear evidence for volcanic activity on the Mercurian surface and none at all for Plate Tectonics. Moreover, the best lunar analogue for the Mercurian smooth plains are *not* the maria, which are the best candidates on the Moon for volcanic activity, but the lunar highland plains, which do not seem to have a volcanic origin. This analogy is made on the basis that the lunar maria are more lightly cratered than the Mercurian smooth plains and have a lower albedo than the surrounding terrain, whereas the lunar highland plains are comparably cratered to the Mercurian smooth plains and like the smooth plains have a similar albedo to their surroundings. (Until the landing of Apollo 16 in a highland plains area it was thought that the lunar highland plains might be of lava origin, but the discovery of copious quantities of breccia indicates that the fairly level surface and the lower crater density than the surrounding highlands is probably because of an abundant fall of ejecta.) The best topographic candidates for volcanic activity of Mercury are a handful of small *rimless* craters and certain ridges that could be solidified lava that once rose from fissures.

It thus seems that there has been very little geological reworking of the Mercurian surface since the apparent planet-wide obliteration of most craters at the beginning.

Planetary interiors are ultimately expected to experience declining temperatures, and there is some fairly direct evidence that to some extent this has already occurred in Mercury. The evidence is provided by a number of low scarps which wind for considerable distances across the Mercurian surface. A segment of one of these is shown in Figure 7.7 (a). Careful scrutiny reveals that the cross-sections of such scarps are something like that shown in Figure 7.7 (b), which in turn can represent the situation shown in Figure 7.7 (c) in which one piece of surface is thrust over another at a low angle. Not surprisingly this is called a *low-angle thrust fault*. By contrast a *lava flow-front* is more likely to have a cross-section something like that in Figure 7.7 (d). Moreover, whereas a lava flow front would have half buried any pre-existing craters a low-angle thrust fault would intersect such craters, as observed on Mercury.

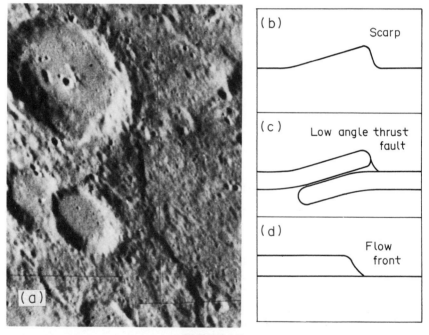

FIGURE 7.7
Discovery Scarp imaged obliquely by Mariner 10, and its low-angle thrust-fault interpretation. The larger crater cut by the scarp is about 60 km across.

Low-angle thrust faults can be produced by the cooling and consequent slight shrinkage of the material beneath, the surface layer then being too large. To produce the low scarps which wind across the Mercurian surface the radius of Mercury need only have decreased by a few *tenths* of a percent. The decrease need not have been distributed uniformly along the radius.

A few sinuous ridges are seen, steep on both sides, and these too may have been raised by interior contraction.

As well as the winding scarps there are also some straight scarps, usually shorter. These may be *strike-slip faults* which occur when solid material suddenly yields under stress as illustrated in Figure 7.8 (a) and (b). There is often some vertical movement along the fracture line, thus yielding a low scarp as illustrated in Figure 7.8 (c). Strike-slip faults could result from interior contraction, though they do not reduce the surface area. However, in the case of Mercury it is likely that most of any strike-slip faults are the result of the tidal slowing by the Sun of Mercury's axial rotation. This would explain why such faults tend to lie in certain directions. Subsequent internal contraction could find such faults a ready site for producing further movement. Also, the short scarps opposite the Caloris impact could be the result of seismic waves acting on such pre-existing faults.

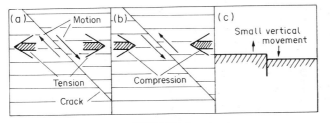

FIGURE 7.8
Strike-slip faults. In (a) and (b) the view is from above and (c) is a vertical section.

7.4 Mercurian volatiles?

The merest traces of various gases exist at the surface of Mercury, amounting to about 10^{12} times *less* than the mass of the Earth's atmosphere and which can be entirely accounted for by capture from the solar wind and by the radioactive decay of certain isotopes in the surface materials.

There is an absence of weathering of any surface features, indicating that Mercury's atmosphereless state stretches back towards the beginning, and there is no evidence for any subsurface volatiles.

Explanations of the likely strong depletion of volatiles in Mercury were outlined in section 5.3.

7.5 The evolution of the interior and the surface of Mercury

One of the earliest events in Mercurian history of which we have any record is the widespread obliteration of craters near the end of accretion, or soon afterwards. This obliteration can be considered in relation to the two possible cratering histories shown in Figure 6.9 for the Moon. The justification for applying this Figure to Mercury with roughly the same time scales will be made in Chapter 8.

If curve *A* in Figure 6.9 applies to Mercury then the erasure of craters must have occurred throughout a fairly extended period during the final stages of accretion. If curve *B* applies then the erasure could have been largely confined to the lull between accretion and the subsequent peak in bombardment, possibly extending into that bombardment. However, it is not at present possible to discern how these histories of cratering and erasure intertwine on Mercury. The survival today on the intercrater plains of largely unaltered patches of that ancient crater-free terrain must owe much, as outlined earlier, to the fairly high surface gravity of Mercury, which confines most ejecta fairly close to the primary impact.

The erasure was probably *not* the result of weathering, because Mercury has probably never possessed a significant atmosphere and certainly not for any length of time. It could have been caused by extensive lava flows or by a more extreme form of the sort of surface plasticity seen in many of the oldest surviving features. In either case powerful sources of heating are required. Such heating could be the result of rapid accretion, or of the presence of appreciable quantities of short-lived quick-acting radioactive isotopes, such as ^{26}Al.

If Mercury was hot in these early times, and it accreted homogeneously, then it would have differentiated early in its history. Otherwise, differentiation could have occurred later if Mercury contained significant quantities of the longer-lived slower-acting radioactive isotopes, notably those of uranium (U), thorium (Th) and potassium (K). It is also possible that Mercury accreted heterogeneously, in which case it need never have become hot enough to differentiate, though it is then more difficult to account for any widespread crater erasure.

For Mercury to have an iron core which is wholly or substantially molten today it must either possess significant quantities of U, Th and K in the core and deep mantle, or tidal heating by the Sun must be particularly powerful, or the solar wind must be efficient at inducing electric currents in the core.

It is certainly possible that Mercury once had a molten iron core which has since partly or wholly solidified. However, solidification of an appreciable fraction of any such core cannot have happened since the Mercurian surface as we now see it began to be created, because the volume shrinkage of iron when it passes from the liquid to the solid phase is so large that the amount of internal contraction of the Mercurian surface would be far in excess of that observed. For the same reason solidification of an appreciable fraction of the mantle can also be ruled out over the same period of history.

The interior cooling recorded by contraction at the surface must have persisted over a long period. The evidence for this is that some scarps cut across several sizeable craters, whereas others are cut *by* sizeable craters, indicating that there are scarps which are older than appreciable numbers of sizeable craters, whereas others are much younger.

You have seen that the smooth plains may well be of non-volcanic origin. They may be composed of impact ejecta, some of which may have been molten. In the case of Caloris some of the ejecta from the impact would have fallen back into the basin because of the fairly high surface gravity. Elsewhere, much of the smooth plains material could have come from large impacts on the so far unseen hemisphere. If mobilization of dust on Mercury occurs electrically, as has been suggested for the Moon (section 6.4.2), then this could help to spread out concentrations of finely divided ejecta.

It is very difficult to place dates on Mercurian events. If the lunar curves in Figure 6.9 can roughly be applied to Mercury then crater erasure took place before about 4000 Ma ago, and most of the present craters, which may also have given rise to the smooth plains, were in place by about 3000 Ma ago, and little has happened on the Mercurian surface since except for the effects of tidal slow-down and interior shrinkage. The most recent events may be impacts which have yielded craters with brights rays (section 6.3.5) which may be no more than a few tens of Ma old.

The apparent lack of much volcanic activity over much of Mercurian history probably results from cooling of the outer mantle, because a planet of the rather small size of Mercury should cool rather rapidly to fairly low temperatures.

7.6 Summary

At its surface Mercury resembles the Moon: atmosphereless, devoid of volatiles, and still bearing ancient craters because of a long history of geological quiescence. This quiescence could be the result of its small size, leading to a fairly rapid cooling to fairly low temperature in the outer mantle. The lack of volatiles could be the result of the various possibilities outlined in section 5.3. However, in its interior Mercury differs from the Moon in that it is far richer in iron, and even has a weak dipole field which is almost certainly of interior origin.

It is a comparatively poorly explored planet, and therefore many questions remain regarding its interior structure, interior temperature and its evolution. Its surface bears some evidence for an early period of crater erasure, and for a long subsequent period of slight internal contraction, presumably the result of cooling.

7.7 Questions

1. How long does it take Mercury's *residual* precession of the perihelion to rotate the semi-major axis of Mercury's orbit around the Sun *once*?
2. Use the arguments outlined in section 7.1 to account for a $^2/_1$ synchronization of a planet in an eccentric orbit around the Sun.

3. The mean density of Mercury indicates that its interior is substantially different from the interiors of the other terrestrial planets. What is this difference?

4. List the types of feasible observations you would make in order to better constrain models of the interior of Mercury.

5. Draw a sequence of about *six* sketches to illustrate the development of the Mercurian surface, and where appropriate indicate possible associated developments of the interior.

6. There are some indications that the interior of Mercury is hot, and other indications that the outer regions have cooled rapidly. What are these indicators?

7. Outline the evidence that Mercury has never had an appreciable atmosphere.

8

CRATER PRODUCTION
ON THE TERRESTRIAL PLANETS

Only for the Moon have cratered regions with widely different number densities had their radiometric ages estimated. This gives a rough indication of the lunar cratering rate over the last 3900 Ma as shown in Figure 8.1 which is based closely on Figure 6.9. In the current absence of radiometric ages for the surface of Mars, Venus and Mercury, it is of considerable interest to see whether the graphs in Figure 8.1 can be applied to these other planets, possibly with modifications, and thus provide a basis for absolute age determination. It is of equal interest to compare the cratering on different planets to elucidate the nature of the small bodies that have caused the cratering.

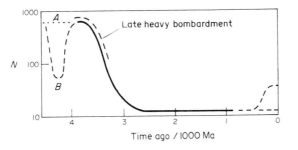

FIGURE 8.1
Impact crater production rates on the Moon (see caption to Figure 6.9).

In comparing craters on different bodies the distinction must be made between the exposure of a planet and the resultant cratering. The *exposure* depends on external factors, in particular on the composition and masses of the population of small bodies that cause cratering, on their number density, and on their orbits as they cross the planet's orbit. The *resultant cratering* depends on the exposure, but also on certain characteristics of the planet, notably the mass of the planet, the surface gravity, the surface composition, and the rate of resurfacing by erosion and by geological processes. You have already seen why some of these factors are important. The importance of some of the others will emerge shortly.

8.1 Recent cratering

Figure 8.2 shows estimates of the *present* rate of cratering of four of the five terrestrial planets. Venus has been excluded because of the paucity of information about impact craters on its surface. The Earth *is* included because though only a few tens of impact craters are known, all of them are less than 500 Ma old and thus constitute recent cratering. Figure 8.3 is a famous but small example. Older craters on the Earth have been erased by weathering and by geological activity.

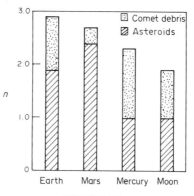

FIGURE 8.2

Present production rates of impact crater, where *n* is the number of craters greater than 10 km diameter produced per 10^6 square kilometres per 100 Ma. Note that the *absolute* values of *n* are accurate only to a factor of 2 up or down but that the *relative* numbers from planet to planet are considerably more accurate.

FIGURE 8.3

Meteor Crater in Arizona. The crater is about 1.2 km across. This is an oblique view.

The estimates in Figure 8.2 are *not* based on observed cratering, which happens far too infrequently, but on the exposure of each planet to the small bodies that can be seen today to move among the terrestrial planets. Broadly speaking there are two types of small body, as indicated in Figure 8.2, asteroids and comet debris. The *asteroids* are small, mainly rocky bodies, most of which occupy orbits that lie between Mars and Jupiter. *Comet debris*, as its name suggests, is derived from comets, and much of this debris lies in orbits that are far more eccentric than most of the asteroids' orbits, and which in many cases carry debris far beyond Jupiter. Comets and asteroids are subjects for Chapter 14.

Craters smaller than 10 km diameter have not been included in Figure 8.2 largely because of the uncertainty about the numbers of the corresponding *very* small bodies that move among the terrestrial planets.

An independent estimate of the recent rate of cratering on the Earth can be made because extensive radiometric dating has been performed of its cratered surfaces. The outcome matches the estimated rate in Figure 8.2, and this provides support for the rates for the other planets. The present rate for the Earth is such that a crater over 10 km diameter is very likely to be produced somewhere on dry land at some time over the next 200 000 years. Don't panic!

The Earth and the Moon are so close together that it has been estimated that the *exposure* of both planets to asteroids and comet debris has been much the same. The estimated difference in the present *cratering rate* shown in Figure 8.2 is largely the result of the different characteristics of each planet.

The exposure of the other terrestrial planets, because they occupy different parts of the Solar System, has been different from that of the Earth and the Moon. For example, though the population of asteroids that can crater Mars today shares some members with the population that can crater the Earth and the Moon, the populations have their own distinct members as well. Moreover, even for the common members the speeds and angles with which they cross the orbit of Mars differ from those at which they cross the orbit of the Earth–Moon system, and this further changes the exposure. When the different characteristics of each planet are also included the outcome is the different present rates in Figure 8.2.

If the *present* cratering rate on Mars can be extended into the distant past then it can be shown that among the light and moderately heavily cratered regions there is a great range of ages, from regions that may be no more than a few tens of Ma old, to regions about 3000 Ma old. The more heavily cratered areas would then be more than 3000 Ma old, regardless of whether the cratering was higher in that distant time. But if the relatively *recent* three-fold increase in cratering rate on the Moon applies also to Mars then many of these Martian ages become older because of the lower cratering rates in the past. However, there is theoretical evidence that the Martian exposure to asteroids was higher *over the last few 1000 Ma* than in more recent times, which scales the ages down such that all but the more heavily cratered terrain is less than about 2000 Ma old.

In the case of Mercury it is estimated that, just as is the case today (Figure 8.2), the slightly higher cratering rate than on the Moon has been a feature of the last few 1000 Ma. If this has been the case then it can be shown that the Mercurian smooth plains are slightly older than the lunar maria. For example, the smooth plains in Caloris would be about 3700 Ma old. This would also be near the date of the Caloris impact, because the infill was probably derived promptly from some of the ejecta.

8.2 The late heavy bombardment

It has been established that on the Moon there was a steep decline in the crater-production rate between about 3900 Ma and 3000 Ma ago. In the early part of this decline the crater production rate was high, and this is called the late heavy bombardment (Figure 8.1), or lhb for short.

Several populations have been suggested as the source of the lhb. Some of these, for example debris within the Earth–Moon system, could not have been shared by the other terrestrial planets, whereas others, for example a general pool of post-accretional debris moving in eccentric orbits through the terrestrial region, would have been shared.

There is certainly evidence that bodies *other* than the Moon suffered lhb: the present cratering rates on the terrestrial planets, when extended back for 4600 Ma, do not produce the crater-saturated regions observed on Mars and Mercury.

The final stages of accretion are expected to be of the nature of an lhb. In the case of the Moon this explanation of the lhb corresponds to graph *A* in Figure 8.1. If this explanation of the lhb applies also to the other terrestrial planets then the degree of sharing of the population producing the lhb would be slight, because in the accretion of planets each planet has access to what to a large extent is a private supply of material.

If instead a graph rather like *B* in Figure 8.1 applies then the prospects of a shared lhb population are enhanced, because a graph like *B* could correspond to the growing post-accretional "transparency" of the Solar System to a population of bodies in highly eccentric orbits traversing the whole terrestrial region and originating largely from beyond this region. From graph *B* it can be shown that the heavily cratered terrain on Mercury extends back to about 4200 Ma, and the intercrater plains would then be older than 4200 Ma. In the case of Mars, if the small bodies had aphelia between Mars and Jupiter then it can be shown that the crater-saturated regions on Mars would be older than 4000 Ma. But if the aphelia lay well beyond Jupiter then it can be shown that these regions would be about 3900

Ma old. (Note that the age of a crater-saturated surface is really the time when the saturation bombardment ended.)

Some evidence for a shared lhb population is provided by the relative numbers of craters of different sizes on the heavily cratered areas of the Moon, Mercury and Mars, making allowances for the influence of different planetary characteristics. These relative numbers indicate that the three planets were exposed to populations of small bodies in which there were the same relative numbers of bodies with various masses. This indicates, but does not prove, that the three populations were one and the same.

The apparent rise in relatively recent times of the cratering rate on the Moon raises the question of whether there could be a new heavy bombardment approaching. This is very unlikely. Over the billenia there has been a general loss of small bodies from the space between the planets, partly to interstellar space and partly through impacts with the planets. Unless matter enters the Solar System from beyond it is hard to see where any large and prolonged increase in exposure could come from. The relatively recent increase could be the result, for example, of a collision between two asteroids which produced several smaller asteroids, or the disruption of a few large comets. In both cases the recent increase may be relatively shortlived.

8.3 Summary

The cratering of the terrestrial planets today is largely the result of exposure to asteroids and comet debris, each planet sharing to a significant extent a common population. The cratering rate on the terrestrial planets was much higher in the distant past, the end of this period being called the late heavy bombardment. It is not clear to what extent the planets were exposed to a common population during this late bombardment.

The graph of the cratering rate for the Moon (Figure 8.1) can only be relied on to give a very general indication of the ages of the various cratered regions on the other terrestrial planets, even if considerable care is taken in adapting the data to these planets.

8.4 Questions

1. Outline the distinction between *exposure* and *resultant cratering* and the factors which make them different from each other. For any given population of small bodies is the *exposure* the same for all terrestrial planets?

2. How many craters greater than 1 km diameter are likely to have been produced on 10^6 square kilometres on the lunar surface between 2000 Ma and 1000 Ma ago?

3. On Mars how many craters greater than 10 km diameter are likely to be produced on a typical 10^6 square kilometre of surface over the next 10^6 years by (i) asteroids, (ii) comet debris? Does the same population threaten the Earth?

4. Outline the evidence that some sort of late heavy bombardment occurred on all terrestrial planets.

5. Outline the arguments for and against the late heavy bombardment of a terrestrial planet being post-accretional.

6. To what extent can the data in Figure 8.1 be applied to the other terrestrial planets?

9

THE JOVIAN SYSTEM

Jupiter orbits the Sun well beyond the orbit of Mars, and though the solar radiation at Jupiter is about 25 times less than at the Earth, Jupiter is so large that only Venus, and Mars at the most favourable oppositions, outshine Jupiter in our skies. Jupiter is about 11 times the diameter and about 1300 times the volume of the Earth, and in its size and in many other respects is a very different world from the terrestrial planets that you have so far met in this book.

Because of its large distance from the Sun Jupiter moves around its orbit rather slowly, and therefore the intervals between oppositions as seen from the Earth (the synodic orbital period) is only about a month longer than a year. Doubtless the slow "stately" motion of Jupiter in its orbit and its brightness were some of the reasons for which the Romans gave to this planet the name Jupiter, the Roman king of the gods.

Galileo was one of the first to observe Jupiter through a telescope, and in 1610 he discovered four satellites which we now call the Galilean satellites. In order from Jupiter they are Io (pronounced as the letters I and O in succession), Europa, Ganymede and Callisto, and are named after four of many of Jupiter's mistresses. Their discovery helped overthrow the medieval concept of the Solar System and a few decades later, by a method that I shall not describe, observations of Io led to the first determination of the speed of light, which until then was thought by many philosophers to be infinite.

Jupiter is now known to have at least sixteen satellites, though the remainder are much smaller than the Galileans. The satellites are listed in Table 9.1. Figure 9.1 shows that on the basis of their orbits they can be divided into three groups. The innermost group consists of eight satellites, including the four Galilean satellites, and it is only the outermost Galilean that is shown in Figure 9.1. The other four lie within the orbit of the innermost Galilean (Io). All eight move in prograde, low eccentricity orbits, of low inclination with respect to Jupiter's equatorial plane. The middle group (Figure 9.1) move in highly inclined and rather eccentric orbits, and the outer group move in rather eccentric, retrograde orbits. The Jovian satellites are the subject of section 9.4.

The two innermost satellites are close to the outer edge of a system of tiny

TABLE 9.1
The satellites of Jupiter

No.	Name	Year of discovery	a/(1000 km)	e	i/°	Sidereal orbital period/days	Radius/km	Density/ (kg/m³)
16	1979J3	1980	127.6	small	small	0.295	about 20	?
14	1979J1	1979	128.4	small	small	0.297	15–20	?
5	Amalthea	1892	181.3	0.003	0.455	0.489	85–135	?
15	1979J2	1980	225.0	small	small	0.678	35–40	?
1	Io	1610	412.6	0.000	0.027	1.769	1816	3550
2	Europa	1610	670.9	0.000	0.468	3.551	1563	3040
3	Ganymede	1610	1070	0.001	0.183	7.155	2638	1930
4	Callisto	1610	1880	0.007	0.253	16.689	2410	1810
13	Leda	1974	11 110	0.147	26.7	240	1–7	?
6	Himalia	1904/5	11 470	0.158	27.6	250.6	85	?
10	Lysithea	1938	11 710	0.120	29.0	260	3–16	?
7	Elara	1904/5	11 740	0.207	24.8	260.1	40	?
12	Ananke	1951	20 700	0.169	147	617	3–14	?
11	Carme	1938	22 350	0.207	163	692	4–20	?
8	Pasiphae	1908	23 300	0.40	147	735	4–23	?
9	Sinope	1914	23 700	0.275	156	758	3–18	?

Jupiter, equatorial radius at 1 bar = 71 400 km.
Primary ring, radius of outer edge = 130 100 km.
Notes: a = semi-major axis of orbit; e = eccentricity of orbit;
i = inclination of orbit with respect to Jovian equatorial plane;
? = not known

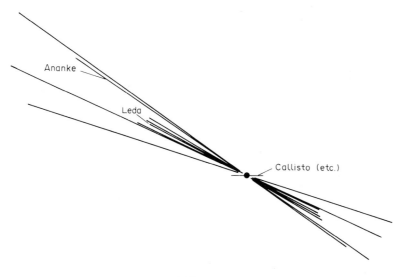

FIGURE 9.1

The orbits of the satellites of Jupiter viewed edgewise to the Jovian equatorial plane (only Callisto of the innermost group is shown). For the sake of clarity the ascending-node longitudes have all been set equal to each other, though within each group they are in any case much the same.

particles that constitute a ring around Jupiter. The existence of this ring had been inferred after the fly-by of Pioneer 11 in 1974, which was the second of four spacecraft that have so far been to Jupiter. But it was not until the fly-by of the next spacecraft, Voyager 1 in 1979, that the ring was directly imaged. Another ring has been discovered, tenuous, and lying between the orbits of Amalthea and Io. Both rings were imaged in greater detail by Voyager 2. The rings will be discussed in Chapter 11.

The four spacecraft missions to Jupiter are listed in Table 9.2. They have furnished a good deal of what we know about Jupiter.

As well as the sixteen satellites and the rings, Jupiter is also surrounded by tenuous clouds of atoms, ions, molecules and electrons. These clouds have a complicated structure, determined largely by the magnetic field of Jupiter and by the interaction of this field with the solar wind. The Jovian magnetic field exerts an influence out to near a surface called the *magnetopause*. The magnetopause is closest to Jupiter on the sunward side but never lies closer to Jupiter than about twice the distance of the orbit of the outermost Galilean satellite. The volume contained by the magnetopause is called the *magnetosphere*, though it is not spherical, the "sphere" implying the *sphere of influence* of the Jovian magnetic field. The magnetosphere is the subject of the next section.

TABLE 9.2
Successful spacecraft missions to Jupiter

Date of Jovian encounter	Name of mission	Type of mission	Comment
4 December 1973	Pioneer 10	Fly-by	Leaving the Solar System
3 December 1974	Pioneer 11	Fly-by	Went on to Saturn
5 March 1979	Voyager 1	Fly-by	Went on to Saturn
9 July 1979	Voyager 2	Fly-by	Went on to Saturn

All missions are from the USA.

The magnetosphere, the rings, the satellites and Jupiter itself, all constitute *the Jovian system*.

9.1 Magnetospheres

Jupiter has the most spectacular magnetosphere in the Solar System. However, several other planets including the Earth are known to have magnetospheres and therefore a general section on such features is appropriate.

● *9.1.1 Planetary magnetospheres*

The magnetosphere is the magnetic "sphere of influence" of a planet, a domain which is limited by the interaction between the magnetic field of the planet and the solar wind.

The *solar wind* originates in the atmosphere of the Sun, and consists of a tenuous gas that has acquired a sufficiently high temperature to escape from the Sun. Indeed, the temperature is so high that no molecules can exist, and most of the atoms are ionized. Because the Sun consists mainly of hydrogen, the solar wind consists largely of ionized hydrogen atoms and electrons, that is, of protons and electrons, though overall the solar wind is neutral. The solar wind is another example of a plasma (section 2.2.2) in this case a plasma in motion.

The outward speed of the solar wind is very high, about 400 km/s, and for various reasons hardly changes as it moves outwards. By contrast, the number density of the ions and electrons falls in proportion to the *square* of the distance from the Sun, because roughly the same number of ions (and electrons) is being spread over a growing spherical area centred on the Sun. At the Earth's orbit the number density of ions is typically 10^{23} times *less* than the atmospheric density at the Earth's surface. The wind is clearly very tenuous, but though it cannot sweep dust before it, it can sweep outwards some of any gas in its path. The solar wind loses its identity in the interstellar medium at the enormous distance of about 100 AU from the Sun.

The number density of the atomic particles in the wind varies on time scales from days to years, in response to variations in solar activity.

The ions and electrons in the wind move *with respect to each other* in such a way that they constitute electric currents which sustain a magnetic field: it is *not* the outward motion of the wind as a whole which generates the field. The origin of the field is, in effect, the magnetic field at the source region of the wind, that is, the surface of the Sun, though this is an extremely complicated phenomenon about which I shall say very little.

When the solar wind encounters the magnetic field of a planet it has the effect of compressing the planetary field on the "upwind" side, and of trailing it out into a long *magnetotail* on the "downward" side. The "upwind" side is illustrated in Figure 9.2, which is the specific case of Jupiter. Beyond the

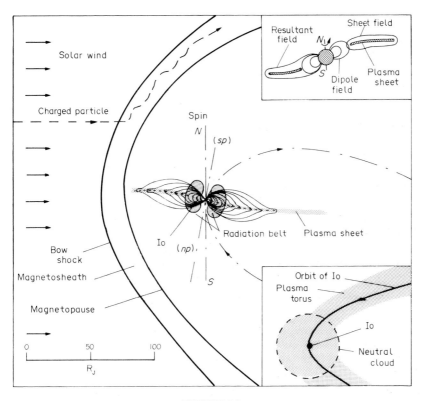

FIGURE 9.2

A cross-section of the Jovian magnetosphere. The viewpoint in the main part of the Figure and the upper inset is such that the orbital plane of Jupiter is seen edgewise. The viewpoint in the lower inset is oblique. Not all magnetic field lines are shown.

surface marked "*bow shock*" interplanetary space has been swept clean of the planetary field by the solar wind. The term "bow shock" arises because of certain similarities between Figure 9.2 and the conditions at the bow of a boat when it is moving through water.

Between the bow shock and the magnetopause the solar wind is greatly decelerated. This leads to highly agitated motions and correspondingly high temperatures. In the case of Jupiter the temperatures are about 300 to 400 million K, though the number density is so low that no glow is visible. The *magnetopause* marks the inner boundary of the agitated region which itself is called the *magnetosheath*. The three-dimensional picture roughly corresponds to rotating the bow shock and magnetopause around the direction along which the solar wind blows outward from the Sun. The space inside the magnetopause is the *magnetosphere*.

In order for a planet to possess a magnetosphere it need not have an internal magnetic dipole moment. Surface magnetization or an ionosphere can also provide a magnetosphere.

The small fraction of the solar wind consisting of neutral atoms readily crosses the magnetopause into the magnetosphere. But most of the ions and electrons are deflected, and of the few which penetrate most do so up the magnetotail. The same barrier exists to other external sources of ions and electrons. Thus, unless there are appreciable sources of particles within the magnetosphere it will contain far less material than the interplanetary space beyond.

Inside the magnetosphere possible sources of particles include the exosphere of the planet, the surfaces or exospheres of any satellites, and the surfaces of any ring particles. These sources yield particles which initially have comparatively low energies of motion. Much more energetic particles can be produced by cosmic-rays, which themselves are so energetic that they can readily pass right through a magnetosphere. However, if a cosmic-ray particle collides with matter inside the magnetosphere neutrons can be ejected from the atoms in the matter. Such neutrons need not themselves be particularly energetic, but in a matter of minutes a free neutron will decay into an energetic proton and electron, though not so energetic as to readily escape from the magnetosphere. Cosmic-rays can also eject atoms, ions and electrons from matter in the magnetosphere. This is called *sputtering*.

Energetic particles can also be produced in collisions between particles, and particles can be accelerated via changes in the position of the magnetopause resulting from variations in the solar wind, and by changes in the magnetic field configuration in the magnetotail. Both of these latter accelerating mechanisms are important, and they are also very complicated. Particles can *lose* energy of motion by colliding with other particles, and by colliding with satellites, rings, or exospheres. Collisions with satellites, rings and exospheres can also result in the loss of particles from the magnetosphere, and also, by sputtering, to the ejection of new ones.

Ions and electrons can recombine to form neutral atoms, and new ionization can result from collisions between particles and from solar uv radiation.

An important repository for ions and electrons is a *radiation belt* which encircles a planet, in which charged particles are fairly effectively trapped by the planet's magnetic field.

The equilibrium number of particles in a magnetosphere depends on the loss and gain rates. Some of the loss processes are determined by the planet's magnetic field such that the stronger the planetary field the slower the overall loss rate and therefore the greater the equilibrium number of particles. Also, the stronger the planetary field the larger the magnetosphere for given solar wind conditions.

There are several other phenomena associated with magnetospheres. For example, leakage of charged particles into the upper atmosphere of a planet can give rise to emissions of em radiation called *aurora*.

9.1.2 *The Jovian magnetosphere*

The main features of the Jovian magnetosphere are shown in Figure 9.2, and have been mostly established by the four spacecraft in Table 9.2.

This is a very large magnetosphere. When the solar wind is weak the upwind magnetopause can lie 100 Jupiter equatorial radii (R_J) from Jupiter. In this case, were it visible from the Earth and if Jupiter were near opposition, the angular diameter of the magnetosphere would be nearly 3 times that of the full Moon. When the solar wind is strong the magnetopause retreats to about 50 R_J upwind. The magnetotail is thought to extend downwind to beyond the orbit of Saturn, and it was detected by Voyager 2 *en route* to Saturn from Jupiter. The large size of the Jovian magnetosphere is partly a consequence of the fairly low particle density in the solar wind at this distance from the Sun, and partly because the Jovian magnetic field is very strong.

The existence of a Jovian magnetic field was established in the 1950s by radio-telescope observations, which detected radio waves that seemed to arise from the motion of charged particles in a magnetic field and were certainly non-thermal (section 2.2.1). Periodic variations in certain of these radio waves were observed, and now that Jupiter is known to have a large internal magnetic dipole moment and that the magnetic axis is inclined with respect to the spin axis, it seems fairly certain that it is the motion of the magnetic axis as Jupiter rotates on its spin axis that causes these periodic variations. Therefore their period should be the axial rotation period of the Jovian interior. The sidereal value is 9h 55m 29.710 s, and this is called the *System III sideral axial period*. Systems I and II will be introduced in section 9.2.

The internal magnetic dipole moment of Jupiter is about 19 000 times that of the Earth, which is a factor of 15 greater per unit volume of planet. The field is inclined at about 10.8° from the spin axis of Jupiter, and the centre of the field is offset from the centre of mass of Jupiter by about 0.10 R_J, mainly along the equatorial plane. The polarity of the field is *opposite* to that of the Earth's field, but in view of the many known past reversals of direction of the Earth's field this is not of much significance.

The form of the Jovian field indicates that the currents which constitute the dipole source are largely confined to between 0.6 R_J and 0.9 R_J. The current per unit volume is several times that in the Earth.

Beyond about 10 R_J the form of the field departs appreciably from that corresponding to a dipole source, as shown in Figure 9.2. This is because the dipole field rapidly decreases with increasing distance from the source and becomes comparable to the fields arising from the ions and electrons in the magnetosphere. Most of these ions and electrons lie in the *belt*, *torus* and *sheet* shown in Figure 9.2.

Ions and electrons enter the Jovian magnetosphere by all of the means outlined in section 9.1.1. A particularly copious source is the volcanoes of Io (see section 9.4.1). These emit sulphurous gases, notably sulphur dioxide (SO_2). By a variety of plausible means much of these gases can escape from Io and become dissociated and ionized. This seems to be the source of the *torus* in the lower inset in Figure 9.2 (not to scale). This is a plasma in which the dominant ions are O^+, S^+ and S^{2+}. Ions leak out of the Io torus and provide the magnetosphere with most of its ions that are *more* massive than H^+ (protons). The number density of ions in the torus varies, presumably because of variations in the volcanic activity of Io.

There is evidence that ions and electrons are exchanged between Io and the Jovian atmosphere, constituting massive electric currents that could help heat Io and thus help maintain its volcanic activity. These currents probably also contribute to the displays of aurora seen in the upper atmosphere of Jupiter's polar regions.

Ions and electrons are also lost from Io by sputtering. This also seems to have produced a cloud of neutral atoms around Io (Figure 9.2) consisting mainly of sodium, potassium and magnesium. Neutral atoms enter this cloud rather than Io's torus because they are not influenced by magnetic forces whereas charged particles are.

The *belt* surrounds Jupiter and contains ions and electrons with comparatively large energies of motion. The number density of energetic ions and electrons is so high that were you or I to cross the belt we would receive 500 times the lethal dose of such particles.

The *sheet* also surrounds Jupiter and arises from the leakage of electrons and ions from the belt, though their large energies of motion are reduced to lower values more typical of the torus. There are two reasons for this leakage.

First, the belt, because it contains matter at a higher density than beyond it, exerts a net outward pressure. Second, the matter in the belt is largely ionized, and therefore tends to be swept around with the dipole field as Jupiter rotates. The electrons and ions thus tend to *co-rotate*, that is, to have sidereal orbital periods equal to the planet's sideral axial period. It can be shown that electrons and ions co-rotating at distances from Jupiter greater than the Keplerian orbit of co-rotation (section 6.1.1) tend to move away from the planet. These two tendencies for electrons and ions to move away from Jupiter are opposed by the Jovian dipole field, which tends to contain the electrons and ions. However, it can be shown that the containment is weakest at the *magnetic* equator, and the outcome is leakage from the belt beyond about 20 R_J.

Thus, a co-rotating sheet of plasma is spun out in the plane of the *magnetic* equator. At greater distances, because of the decline in strength of the dipole field, co-rotation is partially lost and the sheet rotates at a rate between that of co-rotation and the far lower rates required by gravity at these distances from Jupiter. The declining magnetic influence also means that the outer region of the plasma sheet is perpendicular to the spin axis, giving rise to the warp in Figure 9.2.

The motion of the ions and electrons in the sheet is such that it gives rise to a net current around Jupiter. This current generates its own magnetic field which adds to the Jovian dipole field, as shown (not to scale) in the upper inset of Figure 9.2. This produces the observed departures from the dipole field beyond about 10 R_J. The field of the sheet varies in accord with variations in the solar wind, though of course the dipole field does not.

Note that the sheet and the fields in the upper inset of Figure 9.2 correspond to one-half of a sideral axial period of Jupiter different from the situation in the main body of the Figure. Note also that in the main body of the Figure for the sake of clarity only one magnetic field line is shown in the space well beyond the plasma sheet.

Leakage of ions and electrons from the Jovian magnetosphere is partly to Jupiter the rings and the satellites, and partly outwards particularly along the magnetotail. Some of the very energetic ions and electrons which reach the Earth and which we observe as cosmic-rays probably originate from Jupiter. Even now a fast ion may be shooting through you that escaped a few days ago from the Jovian magnetosphere.

9.2 Jupiter's clouds and atmosphere

9.2.1 The face of Jupiter

Figure 9.3 shows a typical picture of Jupiter from a recent spacecraft fly-by. Spacecraft missions have yielded pictures with up to 500 times the

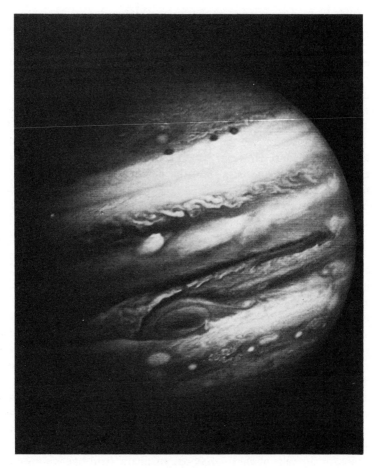

FIGURE 9.3
Jupiter imaged by Voyager 1 from a range of 32.7×10^6 km. North is at the top.

resolution obtained from the Earth. Even so, the larger features in Figure 9.3 are visible from the Earth and many have been observed to hardly change over the 100 or so years for which Jupiter has been subject to extensive and continuous observation with powerful telescopes. In particular, the banded appearance, the positions of the boundaries between them, and even the general colour scheme, have not changed for at least 100 years. However, there have been changes of contrast and changes of detail on time scales from years to days, and for at least 100 years it has been widely realized that we are seeing the tops of richly coloured clouds that cover the entire planet.

The orientation of the spin axis has been determined from the rotation of details in the bands, and it is clear that the bands are parallel to the equator.

The light bands are called *zones* and the dark bands are called *belts*. At latitudes greater than about 40° the banding is less clear, and in the polar regions it is replaced by smaller-scale cellular features.

There is a wealth of small-scale detail within the bands, particularly at the boundaries, and some of this detail is visible in Figure 9.3: there are wavelike features of all sizes and colours, and there are ovals, sometimes called "spots", also of various sizes and colours. These small-scale features are not permanent, though there is a correlation between size and longevity, and the largest of these features, the *Great Red Spot*, is comparatively long lived. This feature is visible in Figure 9.3, and measures 25 000 km by 13 000 km. It was probably first recorded by the British physicist Robert Hooke (1635–1702) in 1664, and more clearly by the Italian astronomer Giovanni Domenico Cassini (1625–1712) in 1665. It has remained at approximately the same latitude since its discovery but it has changed in size, reaching its maximum size of 40 000 × 13 000 km about 100 years ago. By contrast, some of the smallest features seen have lasted for only a few hours.

The wavelike features move relative to each other along the belts and zones, and display a complicated set of interactions. It is also possible to discern rotary motion within the larger ovals, particularly cloud features which circulate clockwise in the northern hemisphere and *anti*clockwise in the southern hemisphere. The smallest ovals of all, which are very numerous, have life cycles which suggest that they too are in rotary motion. However, these same life cycles also suggest that they may have a different nature from the larger ovals. I shall return to this in section 9.2.4.

The east–west drifts of the various small-scale cloud features can be interpreted *either* as wave motions, *or* as mass flows, that is, winds. Powerful evidence *against* the wave interpretation is the *lack* of any correlation between the size of a feature and its rate of east–west motion, a correlation which would exist in the case of waves. Moreover, theories of atmospheric circulation on Jupiter yield winds rather than waves. However, a minority of features may be the result of waves which influence cloud formation as they move through the atmosphere.

The speeds of the east–west winds have been obtained, in effect, by subtracting from the axial sidereal rotation rate of a cloud feature the System III sidereal rotation rate. On the basis of the observations of numerous features in pictures acquired by Voyager 1 the wind-speed graph in Figure 9.4 has been obtained. These are the east–west wind speeds at the cloud tops. The north–south winds are very weak by comparison. The wind speeds did not change appreciably from Voyager 1 to Voyager 2, and certainly seem to be more constant than the small features from which the speeds were obtained. There is some correlation between the wind speeds and the bands, and there are substantial changes in wind speed across belt-zone boundaries which are where most of the small features occur. There is less difference

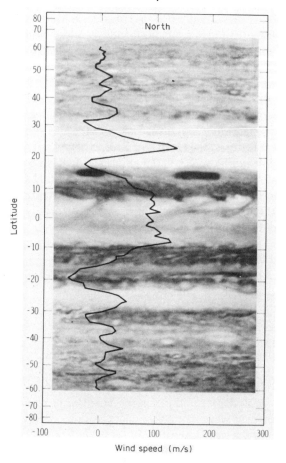

FIGURE 9.4

East–west wind speeds on Jupiter based on Voyager 1 data. Speeds greater than zero correspond to flow from the west to the east.

between the two hemispheres with respect to wind speeds than with respect to the pattern of the bands.

Observations from the Earth of the east–west motion of small features in the Jovian clouds led some time ago to a sidereal rotation period for the equatorial region of 9h 50m 30.003s, and 9h 55m 40.632s for the rest of Jupiter. These are the System I and System II periods respectively. Similar periods can be obtained from the recently observed System III period and the recently observed wind speeds, which indicates the longevity of such winds.

9.2.2 Atmospheric composition and vertical structure

The composition of the atmosphere above the clouds has been largely determined from Earth-based spectrometric studies, starting with the detection of methane and ammonia in the 1930s. By the early 1960s it was established that molecular hydrogen (H_2) is the dominant constituent. Helium (He) has been detected by spacecraft though it has such a weak spectral signature that its abundance has been estimated by indirect means. The relative numbers of molecules in the atmosphere above the clouds is as follows: 90% H_2; 10% He; 0.07% CH_4; 0.02% NH_3; 0.0001% H_2O; plus traces of many other gases. Hydrogen, helium and methane would not condense to form cloud particles at the pressures and temperatures through the cloud regions and therefore their relative abundances apply throughout the atmosphere. Within observational uncertainties these are the same as the solar relative abundances, which are about 0.12 for He/H_2 and about 7 × 10^{-4} for C/H_2, compared with about 0.11 and about 8 × 10^{-4} respectively for the Jovian atmosphere.

Pressures and temperatures in the Jovian atmosphere have largely been determined by spacecraft, using a variety of techniques. The outcome is shown in Figure 9.5. These are globally averaged values of altitude variations, the zero of altitude being taken at a pressure of 1 bar: see section 3.3.1 for the reasons for which a pressure value is an appropriate choice of zero for altitude.

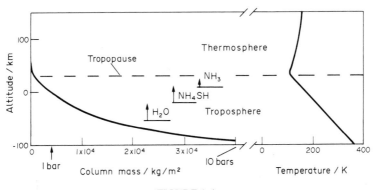

FIGURE 9.5
The vertical structure of Jovian atmosphere.

At altitudes below that at which the pressure is 0.2 bar the lapse rate has the adiabatic value for Jovian conditions. This indicates that thermal convection (section 2.2.2) is the dominant mode of upward heat transfer, and that there are powerful sources of heat deep in the atmosphere. The few percent of solar energy that filters down is just one of these sources — I shall discuss the other sources in section 9.3.3. The dominance of thermal convection is aided by the *infrared opacity* of the atmosphere below 0.2 bar altitude, which reduces the effectiveness of radiation transfer as a competing heat-transfer mechanism. The opacity is largely due to gaseous traces of methane, ammonia, and water. The measurements of pressure and temperature do not penetrate beyond about the 1 bar level, but it seems certain that the adiabatic lapse rate persists to far greater depths. In Figure 9.5 I have continued it down to 10 bars, the pressure versus altitude being calculated from adiabatic lapse rate conditions. It certainly continues beyond 10 bars but this is deep enough to include the main cloud layers that are likely to exist.

The lapse-rate is less than adiabatic above about 30 km altitude, and at about 35 km altitude a minimum temperature of about 105 K is encountered. This indicates the breakdown of thermal convection as an important upward heat-transport mechanism, and by analogy with the Earth's atmosphere the region where it is important is called the troposphere and its upper boundary is called the tropopause. This breakdown above the tropopause arises from the greater infrared transparency of the atmosphere at such altitudes due to the lower atmospheric densities, and because the atmosphere there is appreciably heated by the direct absorption of solar radiation which tends to give an *increase* of temperature with height. Absorption by methane certainly accounts for much of this solar heating, but its abundance may be insufficient to produce the whole effect, in which case there may be an as yet unidentified absorber such as a thin suspension of black soot from the photodissociation of methane.

Temperatures increase with altitude above the tropopause, and from here to where the atmosphere becomes negligible it can be called the thermosphere. The details of the temperature variation with altitude are uncertain above about 150 km altitude, but high in the thermosphere temperatures are typically 850 K and can be as high as 1300 K. Calculations indicate that absorption of solar uv radiation would only raise temperatures in these regions to about 200 K. Clearly, extra sources of heating are required. Upward-moving atmospheric waves could greatly heat a rarefied atmosphere. Alternatively, it could be heated by energetic ions and electrons from the magnetosphere or by some unknown uv absorber.

The temperature at the base of the exosphere is about 1300 K, high enough for there to be an appreciable loss of electrons and ions *to* the magnetosphere, though at comparatively low energies. However, the net *energy* transfer between the magnetosphere and the upper atmosphere is downwards.

High in the thermosphere lies the ionosphere, where the dominant ion is H^+. The ionization is produced by solar uv radiation and by energetic ions and electrons from the magnetosphere.

Several different atmospheric constituents can form clouds in the Jovian atmosphere. A substance can condense to form a cloud if its temperature and its partial pressure lie within the liquid- or solid-phase regions of its phase diagram. You can refer to Figure 2.5, and replace pressure by partial pressure: the partial pressure is the contribution the substance makes to the total pressure, and is proportional to its abundance. If a cloud forms then it will deplete the atmosphere above it of the substances of which the cloud particles are made.

In the Jovian atmosphere the most likely condensates, that is, cloud-making substances, are ammonia, water and a compound of ammonia and hydrogen sulphide (H_2S) called ammonium hydrosulphide (NH_4SH), and though H_2S has not yet been detected in the Jovian atmosphere small amounts should be present. It can be shown that if all the constituent elements in these compounds are present beneath the clouds in solar relative abundances then the *bases* of the three types of cloud will lie at the altitudes shown in Figure 9.5. In all cases the cloud consists of particles in the solid phase. The cloud bases are fairly sharply defined because of the steady decrease of temperature with altitude. For the same reason, any cloud particles which settle downwards quickly evaporate. The convective state of the troposphere tends to raise cloud particles above the base, and at these higher altitudes they do not evaporate because of the lower temperatures, and thus a cloud can have a considerable thickness perhaps extending towards the tropopause, above which convection is weak.

You can see from Figure 9.5 that the uppermost cloud is ammonia, though it could carry traces of other cloud particles borne up from below. It has not proved possible to determine the composition directly, which is difficult for a condensate, but the measured abundance of ammonia in the gaseous phase above the top of these clouds is about what would be expected were this ammonia in equilibrium with solid ammonia at the typical cloud top conditions of 0.5 bar and 148 K. There are thought to be relatively few breaks in the ammonia clouds, though the cloud tops need not be at the same altitude everywhere.

The existence of the ammonium hydrosulphide clouds and water clouds is not firmly established.

Each of the three types of condensate in Figure 9.5 would scatter light uniformly across the visible spectrum, and therefore each would appear *white*. However, the rich variety of Jovian colours could readily be produced by *traces* of other substances in the cloud particles, traces that are far less abundant than the main constituents of such particles. No suitable candidates have yet been detected, but it is readily understood how such traces could arise: solar uv radiation and the thunderstorms which are known to occur in the Jovian atmosphere could act on various atmospheric constituents to produce a rich variety of coloured substances. Indications that such processes are occurring are provided by the existence of molecules such as C_2H_2 and C_6H_6, which can be produced from CH_4 through the action of solar uv radiation or thunderstorms.

Detailed temperature measurements have revealed a correlation between colour and temperature in the Jovian clouds. The highest temperature regions are blue, then comes brown, then white, and then red in the coolest regions of all. Support for the correlation comes, for example, from the observation that white clouds cover and uncover blue and brown clouds. The white clouds are therefore higher, and in the troposphere the greater the altitude the lower the temperature.

The correlation of colour with altitude means that the belts and zones and the smaller-scale features are largely associated with cloud tops that lie at different altitudes. However, the altitude range is only a few kilometres, and therefore with a few exceptions the *major* condensate on view may be ammonia in all regions.

The association of colour with temperature is not in principle surprising, because chemical reactions are very sensitive to temperature. Moreover, the intricacy of the colouring can, and seems to be, influenced by horizontal and vertical circulation patterns in the atmosphere and by waves, both of which can transport material to altitudes where its colours are unstable but where insufficient time has elapsed for chemical reactions to have reached a new equilibrium corresponding to different colours.

But though the general principles controlling the coloration of the Jovian clouds are understood, there is as yet no detailed understanding of how particular features acquire particular colours.

You can see from Figure 9.5 that there exist regions in the Jovian atmosphere where temperatures are about 300 K and the corresponding pressures are a few bars. Given the likelihood of a rich set of different types of molecule, including carbon compounds, the question arises of whether life could have developed and exist today in such regions. However, this seems unlikely, largely because any life-precursing molecules would be carried into regions where they would be destroyed.

The next topic is the circulation of the Jovian atmosphere. But first you need to know a little about the Coriolis effect, named after the French physicist Gaspard Gustave de Coriolis (1792–1843).

● *9.2.3 The Coriolis effect*

In Figure 9.6 the grid lines form an imaginary network of longitude and latitude at a fixed altitude in a planet's atmosphere, the grid lines rotating with the planet. Imagine a "parcel" of atmosphere moving polewards from P and confined by gravity to the altitude of the grid. Its distance from the planet's spin axis must therefore be decreasing. In section 6.5.1 the principle of conservation of angular momentum was outlined. If this principle is applied to the situation in Figure 9.6 then it is clear that as the parcel's distance from the spin axis decreases then the speed of the parcel around the spin axis must *increase*. For example, if at P the parcel had no motion in the

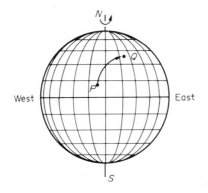

FIGURE 9.6
Illustrating the Coriolis effect.

east–west direction across the planet's surface and the planet's axial rotation is *prograde* then it will pick up motion across the grid towards the east, and in moving northwards will end up for example at Q. If it were in the southern hemisphere and moved southward it would also pick up an eastwards movement. By contrast, movement towards the equator will induce a *westwards* movement across the planet's surface.

Another way in which the parcel can change its distance from the rotation axis is by *vertical* movement, as for example in tropospheric convection, and this too will induce eastward or westward movement.

These east–west movements are examples of the *Coriolis effect*, which in more general terms is the inducement of motion arising from the conservation of angular momentum. It is sometimes called the Coriolis force, but this is very misleading. For example, in Figure 9.6 there is no force acting to push the parcel to the east: the only forces are gravity, which keeps the parcel on the grid, and the force which propels the parcel polewards. From the planet's surface you might *think* there is an eastward force, but there isn't, and if you were *not* rotating with the planet then this would be obvious.

9.2.4 The circulation of the Jovian atmosphere

The Jovian atmosphere has been observed down to about the 1 bar altitude, and throughout this region the predominant motion is east–west. This is because of the rapid axial rotation of Jupiter which yields a powerful Coriolis effect. The motions which feed this east–west trend are probably exemplified by the smallest ovals of all. These seem to be whirls of motion of a particular kind called *eddies*, and at the cloud tops they are very numerous. Figure 9.7 illustrates how they feed the east–west flow. Eddy lifetimes are typically a few days.

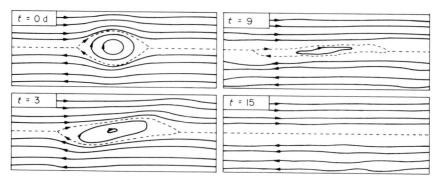

FIGURE 9.7
The transfer of energy of motion from an eddy to an east–west wind. Time t is in days (d).

These eddies seem to be a product of the upward convection in the troposphere, and their energy source is therefore that which drives this convection. Solar energy is one source, but in section 9.3 you shall see that the major source is of internal origin.

Clearly a good deal more solar radiation falls per unit area of the Jovian atmosphere in the equatorial regions than in polar regions, and yet in the observed troposphere there is very little north–south movement. If this were the end of the story then there would be large equator to pole temperature differences in the observed troposphere. But the measurements show that these differences are small, for example only a few Kelvin at the tropopause. Furthermore, the radiation to space at planetary wavelengths is much the same at all latitudes. Clearly therefore equator to pole heat transfer is occurring at greater depths in the atmosphere than those for which wind patterns have been established. However, there is evidence that the east–west flow persists beyond where the atmosphere has been observed, at least for some distance: the stability of the east–west winds would be difficult to understand if only the comparatively small mass of atmosphere above the 1 bar level were involved. I shall discuss circulation at greater depths in section 9.3.3.

It is tempting to regard the larger ovals, perhaps even the GRS, as large eddies. However, their longevity is greater than that predicted by the theory of eddies. Therefore, they are probably not eddies, but though there are several non-eddy models there is as yet no consensus on their nature.

For reasons that are not clear the banding is absent from polar regions, but the generally "granulated" appearance probably arises from upward convection.

In the Earth's atmosphere a far smaller fraction of the atmospheric energy of motion lies in eddies and in any motions to which they give rise than is the case in the Jovian atmosphere. This difference between the two atmospheres is largely a result of the large internal heat source in Jupiter, far more (per unit mass of planet) than in the case of the Earth. However, the transfer of energy of motion from eddies to rapid east–west flow does occur in the Earth's atmosphere, at mid-latitudes and high altitudes. These terrestrial east–west flows are called *jet-streams*.

9.3 The Jovian interior

9.3.1 Observational evidence

The mass of Jupiter was first calculated by Isaac Newton in the seventeenth century from the orbits of the Galilean satellites. This method has been used since, and others have been added to it, including methods based on the effects of Jupiter on the paths of fly-by spacecraft. Jupiter's mass is 317.893 M_E, which is more than 2½ times the mass of *all* the other planets put together.

The equatorial and polar radii have been measured by various means and correspond to some level above the cloud tops. When these values are adjusted to the 1 bar altitude, which lies at a fairly small distance below the cloud tops (Figure 9.5), then the equatorial radius is 71 400 km and the polar radius is 66 550 km. Jupiter is therefore appreciably flattened, though this is not surprising in view of its rapid axial rotation.

The mean density of Jupiter is 1340 kg/m^3. This at once indicates that hydrogen is the dominant constituent of the *whole* of Jupiter. No other substance at the enormous pressures that must exist at even relatively shallow depths in Jupiter could possibly be compressed to so comparatively slight an average density.

The gravitational field has been explored in some detail by observations of the satellites and of the paths of fly-by spacecraft. The departures from a spherically symmetrical gravitational field are so big that J_2 is large enough to have been measured very accurately, and for the same reason this is also the case for the next gravitational coefficient in the series, J_4 (jay-four). The variation in strength of the gravitational field at a constant distance from the centre of mass of a planet associated with J_4 is shown in Figure 9.8. This Figure is drawn in a similar manner to J_2 in Figure 3.5 (b), except that in Figure 9.8 only one quadrant is shown. The other three quadrants are the same, as for J_2. J_4 and J_2 are also similar in that the variation is the same at all longitudes. They differ in that whereas the component of the gravitational field associated with J_2 has *one* zero per quadrant, the component associated with J_4 has *two*.

For Jupiter J_4 is sufficiently large that in modelling the density variation with depth it must be included along with J_2, and the methods outlined in section 3.2.1 have to be modified. Recent models of Jupiter have been

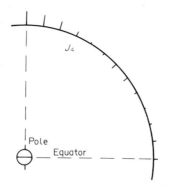

FIGURE 9.8
The $\mathcal{J}_4$ component of the gravitational field.

constrained by $\mathcal{J}_2$, $\mathcal{J}_4$ and the axial period, rather than by the polar moment of inertia. These models assume hydrostatic equilibrium. This assumption can be tested by calculating the flattening from $\mathcal{J}_2$, $\mathcal{J}_4$ and the System III period, assuming hydrostatic equilibrium. The calculated value agrees with the observed flattening to within observational uncertainties, indicating that Jupiter is in hydrostatic equilibrium, and further support for this conclusion comes from smaller-scale details of the Jovian gravity field. This result is no surprise for a planet of Jupiter's mass. The large values of $\mathcal{J}_2$ and $\mathcal{J}_4$, like that of f, arise from the rapid spin of a planet in hydrostatic equilibrium.

The powerful magnetic field of Jupiter indicates that a significant fraction of the Jovian interior consists of an electrical conductor in the liquid phase. Evidence that much of the interior is liquid also comes from observed global vibrations of Jupiter, but I shall say no more about this. I shall return to the nature of the conductor in the next section.

As you shall see in the next section, the predominance of hydrogen in Jupiter means that it does not *have* to have high interior temperatures to be liquid. There is, however, separate evidence which shows that the Jovian interior is indeed hot, namely, that Jupiter radiates energy to space at a greater rate than that at which it absorbs energy from the Sun. The energy radiated to space has been calculated from detailed observations across the Jovian disc of radiation at planetary wavelengths, and the energy absorbed from the Sun has been calculated from the Bond albedo, which measurements have shown is about 35%. It is clear that Jupiter radiates energy to space at a rate somewhere between 1½ and 2½ times the rate at which it absorbs energy from the Sun. In order to explain this *excess radiation* additional inputs of energy to Jupiter from beyond the planet have been considered, but none seem anywhere near large enough to make up the difference. It has also been suggested that the Bond albedo was lower within the last few hundred years and that the atmosphere is still cooling as Jupiter adjusts to its new higher albedo. But there is no independent evidence for this, and even if the albedo in the past was zero, any change would probably be unable to account for the present excess radiation.

Therefore, it seems likely that Jupiter has powerful *internal* sources of heat, and to account for the excess radiation these would make the interior hot. I shall consider these internal sources in section 9.3.3.

Further evidence for a hot Jovian interior is provided by the satellites, as you shall see in section 9.4.4.

Because hydrogen accounts for most of the mass of Jupiter, and, as you shall see, of Saturn, it is very important to consider the behaviour of hydrogen within such planets.

9.3.2 The behaviour of hydrogen in the Jovian interior

It is likely that several per cent of Jupiter consists of helium. However, the story can be simplified without losing its essence by neglecting the helium for the time being.

Figure 9.9 shows the measured phase-diagram of molecular hydrogen. At

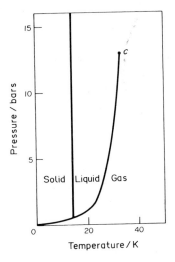

FIGURE 9.9
The phase diagram of molecular hydrogen (H_2).

the greatest depth in Jupiter penetrated by direct measurements the temperature is about 170 K and is rising along the adiabatic lapse rate. This temperature is already well to the right of the point marked c in Figure 9.9, which is the *critical-point*. The temperatures deeper into Jupiter will certainly not fall below 170 K. Thus, although the measured value of 170 K is only at a pressure of 1 bar, the pressure–temperature path on the molecular hydrogen phase diagram as we descend to higher pressures never crosses the line ending at the critical point. This line divides the pressure–temperature conditions for the liquid phase from those for the gaseous phase. If the path had crossed this line then a clear surface would have been encountered some distance below the clouds. Above this surface the hydrogen would have been gaseous, and below it the hydrogen would have been liquid: Jupiter would have thus been covered in hydrogen oceans. Instead, as we go deeper and deeper the gaseous hydrogen *gradually* gets denser and denser until imperceptibly, and at pressures considerably in excess of those in Figure 9.9, it acquires a density more characteristic of a liquid than that of a gas. Therefore, as we descend into Jupiter we encounter no surface, but gradually find ourselves amidst an ocean.

The question arises of whether a *solid* surface of hydrogen is encountered in the ocean. In Figure 9.9 this would occur if the boundary between the solid and liquid phases were to move to the right as pressures get large. Laboratory studies have been made up to about 10^6 bars, at which pressures molecular hydrogen is still liquid even at 170 K. However, such pressures are encountered at depths of only a few 1000 km below the clouds, and therefore theoretical studies have to be used to obtain the phase diagram at greater pressures. These studies confidently predict that at any plausible temperatures in Jupiter no solid molecular hydrogen surface is encountered. These studies also predict that at a pressure somewhere between 2×10^6 bars and 5×10^6 bars an interface is encountered with liquid molecular hydrogen above and liquid *metallic hydrogen* beneath.

Metallic hydrogen is expected to be produced at very high pressures because the molecules of molecular hydrogen (H_2) will be forced so close together that each of the two hydrogen atoms in the H_2 molecule is attracted to atoms in neighbouring molecules as much as to its partner. At this pressure the H_2 molecules break up. Moreover, the single electron that orbits the nucleus of each hydrogen atom will also become equally attracted to neighbouring atoms, and so the atoms break up too, and the hydrogen will consist of a "gas" of electrons moving among the hydrogen atom nuclei. Many of the characteristic properties of metals arise from the existence in them of such "electron gases", and therefore the term "metallic hydrogen" is appropriate.

One of the properties that an electron gas gives to metallic hydrogen, and to most other metals, is high electrical conductivity. Thus, electric currents in liquid metallic hydrogen are an obvious candidate for the source of Jupiter's large internal magnetic dipole moment. Moreover, the transition pressure is encountered somewhere between 0.7 and 0.85 R_J, and the location of the dipole source at no great depth is required by the external magnetic field measurements (section 9.1.2). However, a significant contribution to the internal magnetic dipole moment from liquid iron and other materials nearer the centre of Jupiter cannot be ruled out.

9.3.3 Internal heat sources for Jupiter

For a massive planet consisting largely of hydrogen and helium there are two main sources of heat for the interior, each of which could account for the excess radiation.

First, there is heat left over from the accretion of Jupiter. For almost any likely means of planetary accretion, Jupiter is of such large mass that initially the internal temperatures would have been very high. During the first few tens of Ma Jupiter would cool so quickly that it hardly matters what the initial temperatures were provided that they were at least a few times 10 000 K. After a few hundred Ma, the exact time depending on the initial temperatures but in any case a small fraction of Jupiter's lifetime, the rate of cooling would be far less. By this stage the outer regions of the atmosphere would essentially be controlling the radiation rate, as follows.

In the hot, liquid interior, thermal convection will be the dominant mechanism by which heat is transferred outwards. This is a very efficient heat-transfer mechanism, and if it existed all the way into the region of the atmosphere which radiated to space then by today the Jovian interior would have lost almost all of its heat of accretion, and any remnant could not be a significant source of internal energy. However, as you can see from Figure 9.5, the adiabatic lapse-rate, which indicates heat transfer by thermal convection, is not followed at altitudes above about 30 km, which is below the altitudes from which most of the radiation to space (at planetary wavelengths) originates. Thus, in effect the Jovian interior is "wrapped" in an insulating blanket (compare section 2.1.16). This blanket could have retained a substantial fraction of the heat of accretion thus maintaining sufficiently high internal temperatures today to account for the present excess radiation to space.

The rate of loss of heat of accretion can be reduced even further if the absorption of solar radiation in the atmosphere occurs mainly in the troposphere and largely occurs *below* where the radiation to space originates. If both these conditions are satisfied then the solar radiation can to some extent provide the upward flow of heat by convection, thus reducing the outflow from greater depths. I shall call this effect *solar choking* of the flow. In the case of Jupiter it has probably had only a small effect on the rate of loss of heat. However, the greater solar input at the equator than at the poles would produce greater solar choking at the equator, and this can be shown to aid equator to pole circulation in the Jovian interior, thus distributing solar energy more uniformly and leading to the small equator to pole temperature differences observed in the upper troposphere.

The *second* main source of internal energy is heat from *gravitational separation*. You have already met an example of this in the heat generated when the iron core of a planet separates downwards. The source of energy is gravitational. In the case of Jupiter the downward separation of *helium* has to be considered. In any plausible way of accreting Jupiter the hydrogen and helium are initially well mixed at a molecular level. Moreover, at all plausible temperatures for the Jovian interior then or now any mixing on a *molecular* scale is maintained even in the liquid phase. This is because, though the helium atom is twice as massive as the H_2 molecule, the random collisions between atoms and molecules overwhelm the tendency for gravitational separation. However, if *droplets* of helium form then the random collisions of atoms and molecules on the droplet largely cancel out whilst the tendency for grativational separation rises in proportion to the number of atoms in the droplet. Thus the droplets may separate. However, one further condition must be fulfilled, namely that the upward speed of convective motions must not be sufficient to bear the droplet aloft. In the case of Jupiter this condition is probably fulfilled, and therefore the gravitational separation of helium hinges on the extent to which droplets form, that is on the degree to which helium is *miscible* in hydrogen, and this depends on the pressure and the temperature.

Gravitational energy can be released in a different way by the growth of the metallic hydrogen region at the expense of the molecular hydrogen region, because metallic hydrogen at comparable pressures and temperatures is the denser. This interacts with helium separation because helium is probably more miscible in molecular hydrogen than in metallic hydrogen, and therefore the growth of the metallic phase can enhance the rate of gravitational separation of helium.

Gravitational separation of helium from hydrogen is a self-regulating rather than an unstable process: the higher the temperature the greater the miscibility of helium and planetary temperatures increase towards the centre.

The miscibilies of helium in molecular and metallic hydrogen at temperatures and pressures appropriate to Jupiter are very uncertain, and therefore the importance of this source of internal energy in the past and today is uncertain. However, it seems plausible that this source of energy has been of little importance in the *past* because Jupiter is so very massive that it probably formed very hot, though Jupiter may *now* be sufficiently cool for helium separation to have recently become significant, or for it to become significant in the relatively near future.

9.3.4 Models of the Jovian interior

You have seen that there exists convincing evidence that the Jovian interior is hot and that convection is the dominant mode of heat transport outwards. Therefore the lapse-rate is adiabatic to very great depths. Most of the recent models of Jupiter have incorporated this feature.

A typical approach is to assume a ratio of hydrogen to helium which is the same or fairly similar to that measured in the atmosphere, and to assume that there is a small fraction of heavier elements and their compounds which may be concentrated in a core or dispersed to a greater or lesser extent. The helium may also be separated to some degree. The temperature at a pressure of 1 bar is taken as a reference point from which, using the adiabatic lapse-rates of the particular mix of materials at each depth and their equations of state (section 2.1.4), calculations are made of the temperatures pressures and densities at greater depths. The details of the model are adjusted until it can simultaneously yield values acceptably close to the observed values of: the equatorial radius (usually at 1 bar); the mean density; the axial rotation rate; J_2; J_4; and the radiation excess. The model must also be consistent with the existence of a magnetic dipole source at no great depth, have plausible relative abundances of the various elements, and in representing Jupiter today can be the outcome of a plausible evolution from the birth of Jupiter to the present.

Figure 9.10 shows the main features of one such model of Jupiter that satisfies the various constraints. The gravitational separation of helium is negligible and in this model always would have been, because the evolutionary sequence that yields this *present* state has a hot start from accretion 4600 Ma ago, the atmospheric regulated loss of the heat of accretion bringing Jupiter to its present state. However, it is possible that Jupiter

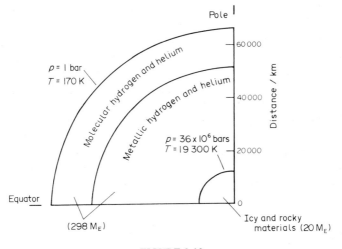

FIGURE 9.10
A model of Jupiter.

resembles Figure 9.10 today *without* having followed this particular evolution.

The most certain features of Figure 9.10 are that Jupiter consists largely of hydrogen, that much of the interior is liquid/gas, that there is no atmosphere/surface interface, that there is some concentration towards the centre of icy and rocky substances (this is required by the density versus depth), and that the hydrogen exists in metallic form beyond a fairly modest depth.

At a more detailed level considerable variations are possible from model to model, such as the exact ratios of hydrogen to helium and of hydrogen to the other substances.

Many of the differences between the various models arise from limited knowledge of the miscibilities of helium and other substances in hydrogen, and from limited knowledge of the equations of state of the various substances. The equation of state of molecular hydrogen at high pressures and temperatures is particularly important and yet it is poorly known. Moreover, the simplification cannot be made, as it can in the terrestrial planets, that the influence of temperature on density is slight compared with the influence of pressure: the temperature change with depth in adiabatic lapse-rate models of Jupiter is much too large for this to be the case.

Nevertheless, a broad understanding of the present Jovian interior, and how it could have got that way after its initial accretion, now exists. The initial accretion itself is a subject for Chapter 15.

9.4 The Jovian satellites

I shall be mainly concerned with the four Galilean satellites, which, as you can see from Figure 1.2 and from Table 9.1, are comparable in size to the terrestrial planets. The other satellites of Jupiter are far smaller.

9.4.1 The Galilean surfaces

Figure 9.11 shows the surfaces of all four Galileans. None of these satellites have appreciable atmospheres, and they are all in synchronous rotation with respect to Jupiter. Infrared spectrometry, mainly Earth-based, has revealed that water ice is an abundant constituent of the surfaces of all the Galileans, except Io.

The surface of *Callisto*, the outermost Galilean, is very heavily cratered, close to saturation over much of its surface. The topography of the craters shows that they are the result of impact, and that the surface consists largely of water ice rather than of rock. Other common ices, particularly methane, ammonia and carbon dioxide, can be ruled out because surface temperatures on Callisto reach 140 K, which would lead to their rapid evaporation and loss

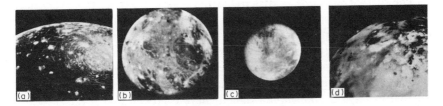

FIGURE 9.11
The Galilean satellites of Jupiter imaged by Voyager 1 (not shown to their correct relative sizes). (a) Callisto, (b) Ganymede, (c) Europa, (d) Io.

to space. However, Callisto is too dark for the surface to consist of water ice alone. Moreover, some cratering has exposed a white powder, which suggests that the other components are largely confined to the surface. Furthermore, in the spectra of surface materials the signatures of water ice are rather subdued. This could be caused by a separate rocky component, or by the presence of hydrated silicates.

Parts of the surface of *Ganymede* are as heavily cratered and as dark as Callisto. But elsewhere it is less heavily cratered and less dark, and covered with sets of grooves that are grouped into bands which exhibit intricate patterns. This suggests that the surface of Ganymede has been reworked to a greater extent or more recently than that of Callisto. These two satellites have rather similar masses and densities, and perhaps therefore it is their different distances from Jupiter that has led directly to these surface differences. However, the reasons are not at all clear.

The surface of *Europa* is remarkably smooth: there are very few impact craters, and the numerous ridges a few kilometres in width which cross the surface are only a few hundred metres high. There are also numerous dark stripes, tens to hundreds of kilometres across, but of negligible height or depth. These ridges and dark stripes can be explained by the freezing of a shallow ocean of liquid water that covered the globe beneath a thin surface of water ice. As it froze the surface would be extensively cracked, and the cracks would be filled from beneath, producing dark stripes, their lower albedo resulting from greater concentrations of impurities than in the original icy crust: the high albedo of the original crust that today lies between the dark stripes shows it to be fairly pure. The freezing also produced the ridges, by crumpling the *original* crust, the ridges having a similar albedo to their surroundings. The scarcity of impact craters would result if the liquid ocean persisted for some time after the origin of Europa. The heat to sustain such an ocean could have come from radioactive decay: the density of Europa is sufficiently high for it to contain sufficient silicates that, in turn, could have borne sufficient radioactive isotopes to produce the heating. Callisto and Ganymede are less dense, suggesting less radioactive heating, which could be why they have such different surfaces from Europa.

The surface of *Io* differs strikingly from that of the other three Galileans. There are no impact craters at all, and nearly all the features are clearly of volcanic origin. Indeed, there are several active volcanoes visible today one of which can be seen at the edge of Io in Figure 9.11. No water, or any other ices, have been detected at the surface of Io, and even if there ever were any such ices the interior heating indicated by the volcanism would have driven them off. Sulphur and sulphur dioxide have been detected at the surface in abundance, and they can readily account for the richly coloured surface of Io. It is clear that much of this sulphurous material comes from the volcanoes, though it is not known whether it is the main constituent of the erupted material, or whether it is a heavy contaminant of, for example, molten silicates.

The volcanic activity on Io seems to be maintained by tidal forces, as follows. Calculations show that Europa, and to a smaller extent Ganymede, produce a slight eccentricity in the orbit of Io which otherwise would be very nearly perfectly circular. This slight eccentricity means that the tidal bulge on Io, which because of the synchronous rotation would otherwise point straight at Jupiter, swings to each side of this line as Io goes around its orbit. The constant tendency of Jupiter to pull this bulge into line will cause internal heating as the bulge moves under Jovian influence. The heating effect is greater if the interior is liquid, and if the tidal heating is to account for the volcanic activity of Io it seems necessary for the interior of Io to be liquid: the tidal heating probably could not melt a solid interior but it can *maintain* a liquid phase.

A supplementary source of internal heating is through the electric currents that flow between Io and Jupiter, as outlined in section 9.1.2. These currents could also aid the observed ejection to high altitudes of material from volcanic eruptions on Io.

Absolute ages for the Galilean surfaces can be roughly estimated from the number of densities of impact craters. This is taken up in section 10.5.

9.4.2 The Galilean interiors

Figure 9.12 shows plausible models of the Galilean interiors. These models are based on the mean densities, obtained from observations, and on the likelihood that, even if the accretion was homogeneous, the interior temperatures managed to reach the modest values required for gravitational separation of the denser rocky substances from water. Note that surface temperatures today are high enough to rule out ices other than water ice as abundant constituents.

On the basis of their mean densities alone (Figure 9.12) it is clear that Io and Europa consist largely of materials with densities characteristic of

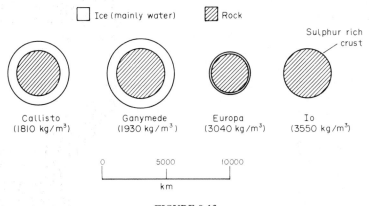

FIGURE 9.12
Models of the Galilean interiors.

silicates, whereas Ganymede and Callisto must contain substantial amounts of water ice. Therefore, in spite of their comparable sizes to terrestrial planets, Ganymede and Callisto are not at all terrestrial in their interiors.

The nature of the rocky component is not certain, though silicates are likely to be abundant, and the higher density of Io suggests that iron-rich compounds, perhaps iron itself, may be present in significant amounts.

9.4.3 *The other Jovian satellites*

The Galileans are by far the largest members of the innermost of the three groups of Jovian satellites (Figure 9.1). Of the four remaining satellites in this group (Table 9.1), the tiny satellites 14, 15 and 16 (they have no names) may play a role in determining the structure of Jupiter's rings, as outlined in section 11.1. The remaining satellite *Amalthea*, is rather larger than satellites 14, 15, 16. It has an irregular shape with its largest dimension of 270 km pointing towards Jupiter, presumably through the action of tidal forces. It has a low albedo and is tinted red, possibly by sulphur from Io. Its density is unknown, but if it is close to that of water ice then it is within the Roche limit of Jupiter for a body with such a density (see section 6.1.1). However, it could remain undisrupted because of its rather small size, non-gravitational cohesion still being of importance. If it is rocky then it lies outside the Roche limit.

The other two groups of satellites consist of very small bodies (Table 9.1) moving in unusual orbits (Figure 9.1).

9.4.4 *The origin of the Jovian satellites*

It is likely that both of the outer groups of satellites have been captured by

Jupiter. Indeed, the outer group would not have to be much further from Jupiter to break free and orbit the Sun separately. Further evidence for their capture origin is a broad resemblance between the albedo and spectra of many of these satellites and the albedo and spectra of CR type asteroids to be outlined in section 14.4. This suggests that each group is a disrupted CR type asteroid. It is not certain how Jupiter could capture any such asteroids today. However, in the distant past capture would have been aided by the presence of any residual gases around Jupiter, since dispersed. Moreover, such gases would have been particularly effective at capturing bodies of about the right size to account for these two groups. Bodies much larger would not have been captured, and bodies much smaller would have gradually fallen into Jupiter.

It is *very* hard to see how these two groups of satellites could have formed as part of the Jovian system and end up with such far-flung, eccentric, highly inclined, even retrograde, orbits.

By contrast, the proximity of the inner group to Jupiter suggests that all or most of them were formed as part of the Jovian system. It would certainly be very difficult to capture them in such close orbits. (Note that the low inclinations and small eccentricities would readily evolve from any larger values in such close orbits.) Moreover, if, as seems likely, Jupiter was very hot after accretion or subsequently passed through a brief hot phase because of helium separation, then water could have been evaporated more readily from rocky bodies near Jupiter than from those further away, thus providing an explanation of the lower abundance of ice in Io and Europa than in Ganymede and Callisto.

9.5 Summary

Jupiter is very different from the terrestrial planets. It is far more massive and consists largely of hydrogen. Indeed the relative abundances of the elements in Jupiter are probably fairly similar to solar relative abundances. This is no surface below the clouds, and the interior is probably at far higher temperatures than in many of the terrestrial planets, largely as a result of heat left over from accretion.

The internal source of heat, plus the rapid axial spin, leads to patterns of atmospheric circulation different from those of the terrestrial planets. The visible clouds probably consist of solid particles of ammonia coloured by unknown substances in a manner that is poorly understood.

Jupiter has a modest system of rings and many satellites, four of which are of terrestrial planet size. The large magnetic dipole moment of Jupiter has resulted in an enormous magnetosphere, which contains a rich variety of plasma. The magnetic dipole moment probably originates, in the main, in a shell of liquid metallic hydrogen.

9.6 Questions

1. In a sentence state how a plasma can generate a magnetic field.
2. What are the problems with making and using a spacecraft to "sail" in the solar wind?
3. Why does Jupiter have a large and variable magnetosphere?
4. If you travelled straight out from the Sun through the centre of Jupiter then in what order would you first meet the following things: a magnetic field dominated by Jupiter's internal dipole source; Io; bow shock; radiation belt; plasma disc; magnetopause; magnetotail; plasma torus, rings; magnetosheath.
5. List the sources and loss-mechanisms of ions and electrons in the Jovian magnetosphere and their acceleration and deceleration mechanisms.
6. A planetary atmosphere can exist in which there is *no* appreciable internal (non-solar) source of heat, and yet in which an adiabatic lapse rate exists *below* where nearly all solar energy is absorbed. Name one such planet, and explain how this situation arises? (It's not Jupiter.)
7. A planet spins in the retrograde direction, and a parcel of atmosphere at a fixed altitude moves northwards from the equator. Does it acquire an eastward or a westward drift? If it rose *vertically* at a fixed latitude then which drift direction would it acquire?
8. Outline the factors which influence the circulation of the Jovian atmosphere, making links to associated cloud features wherever possible.
9. List the factors which influence the rate of loss of heat by the Jovian interior.
10. Contrast the types of heat source available to the Jovian interior with those available to the Earth's interior. Which of the possible Jovian sources are likely to have been significant in the past?
11. List the observational and theoretical constraints on constructing models of the present Jovian interior, and outline why there exists a range of plausible models. Which of the constraints apply also to models of the *evolution* of Jupiter?
12. Discuss whether the *Galilean* satellites should be divided into two or into three groups.

10

THE SATURNIAN SYSTEM

Saturn, because of its greater distance from the Sun, creeps along its orbit more slowly than Jupiter and moreover it is dimmer and yellower. These attributes seem to have been suggestive of the god of time to the ancients: an old, yellowed man who looks after the ages and reaps all things mortal. It's the Roman name for the god of time by which we know Saturn.

Saturn is smaller, but not a lot smaller, than Jupiter, about 80% of Jupiter's diameter. And in many other ways the two planets are broadly similar.

But there is a well-known respect in which the Saturnian and Jovian systems differ, namely, that Saturn possesses a vast and spectacular system of rings. Galileo in 1610 noticed that Saturn had some sort of appendages, but it was not until 1659 that it was firmly established by Huyghens that the appendages were

". . . a flat ring which nowhere touches the body of the planet . . ."

This single ring has since been shown to consist of two major rings, A and B. Several other rings, labelled C to G, have also been discovered. These major rings, A to G, are indicated in the lower part of Figure 10.1. This Figure also shows that the ring system is very extended, though it is also very thin. Figure 10.1 does no justice to the intricate structure of the ring system: in part this will be put right in Chapter 11.

Four years earlier, in 1655, Huyghens had discovered a *satellite* of Saturn, the first planetary satellite to have been discovered since the Galileans of Jupiter in 1610. It was named *Titan*, which, in mythology is the generic name for the giant children of Uranus and Gaia. Titan is indeed a giant, second in radius among planetary satellites only to Ganymede, and larger than the planet Mercury. However, because of its lower density Titan like Ganymede is not as massive as Mercury. Titan is the only satellite known to possess an appreciable atmosphere.

Saturn is now known to have at least seventeen satellites, and these are listed in Table 10.1. The eight named satellites discovered since Titan owe their names to a suggestion made early in the nineteenth century by the British scientist John Frederick William Herschel (1792–1871), who

240

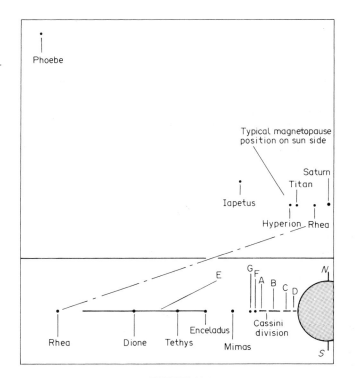

FIGURE 10.1
The rings of Saturn and the orbits of the satellites viewed edgewise to the equatorial plane of Saturn (the small satellites have been omitted). Note that the orbits are shown in cross-section and thus appear as dots, and that, for the sake of clarity, the ascending-node longitudes have all been set equal to each other.

suggested that the satellites of Saturn should have names derived from the Titans. The eight *un*-named satellites have all been discovered in recent years, largely by means of the three spacecraft missions to Saturn. These are listed in Table 10.2, and it is these missions that have furnished a good deal of what is known about the Saturnian system.

From Table 10.1 you can see that satellites 10 and 11 have very similar orbits. At the moment they are on almost opposite sides of Saturn, but because their orbits are not quite identical 11 is slowly catching up with 10. Moreover, the separation between their orbits is less than the sum of the radii of the satellites and therefore a collision might seem certain. However, a careful analysis has revealed that as they get close they will gravitationally influence each other's orbits in such a way that the satellites will then move apart. This sequence of events will continually repeat itself, the net effect being that relative to one of the pair the other executes a horseshoe-shaped path.

TABLE 10.1
The satellites of Saturn

No.	Name	Year of discovery	a/(1000 km)	e	i/°	Sidereal orbital period/days	Radius/km	Density/ kg m³
15	1980S28	1980	137.7	0.002	0.30	0.602	30	?
14	1980S27	1980	139.4	0.003	0.00	0.613	70×40	?
13	1980S26	1980	141.7	0.004	0.05	0.629	50×30	?
10	1980S1	1980	151.4	0.007	0.14	0.694	110×80	?
11	1980S3	1980	151.5	0.009	0.34	0.695	70×50	?
1	Mimas	1789	185.6	0.020	1.517	0.942	195	1200
2	Enceladus	1789	238.1	0.004	0.023	1.369	255	1100
3	Tethys	1684	294.7	0.000	1.093	1.885	525	1000
	1980S25	1980	294.7	?	?	1.885	25	?
	1980S13	1980	294.7	?	?	1.885	30	?
4	Dione	1684	375.5	0.002	0.023	2.733	560	1400
	1980S6	1980	378.1	0.005	0.15	2.739	30	?
5	Rhea	1672	527.2	0.001	0.35	4.511	765	1300
6	Titan	1655	1222	0.029	0.33	15.91	2575★	1900
7	Hyperion	1848	1483	0.104	0.4	21.28	205×110	?
8	Iapetus	1671	3560	0.028	14.7	79.15	720	1200
9	Phoebe	1898	12950	0.163	150	549.1	100	?

Saturn, equatorial radius at 1 bar = 60 000 km.
A ring, radius of outer edge = 136 200 km.
Notes: a = semi-major axis of orbit; e = eccentricity of orbit;
i inclination of orbit with respect to Saturn's equatorial plane;
★ radius of solid surface of Titan, and the corresponding density;
? = not known.

TABLE 10.2
Successful spacecraft missions to Saturn

Date of Saturnian encounter	Name of mission	Type of mission	Comment
1 September 1979	Pioneer Saturn	Fly-by	Pioneer 11 renamed
12 November 1980	Voyager 1	Fly-by	Leaving the Solar System
28 August 1981	Voyager 2	Fly-by	Going on to Uranus (January 1986) and then Neptune (September 1989)

The satellites 1980S6 and Dione also occupy very similar orbits, the one lying about 60° around the orbit from the other. A similar situation exists for Tethys and for the two small satellites 1980S13 and 1980S25, one of which is 60° *ahead* of Tethys, the other 60° *behind*. These 60° positions are two of several *Lagrangian points*, which are places where a small body can be gravitationally stable with respect to two much more massive bodies. In the case of 1980S6 Saturn is one of the two massive bodies, Dione being the other. In the case of 1980S13 and 1980S25 Tethys substitutes for Dione. Note that the 60° Lagrangian Points are stable positions for the small satellites, and that no such positions exist for the satellites 10 and 11 because there is no massive satellite near their orbits. The Lagrangian points are named after the Franco-Italian mathematician Joseph Louis Lagrange (1736–1813).

With the exception of Phoebe ("fee-be"), the satellites lie fairly close to Saturn, at distances comparable to those of the inner group of Jovian satellites. A cross-section of the orbits of the *named* satellites is shown in Figure 10.1. The un-named satellites all lie within the orbit of Rhea. Except for Phoebe and Iapetus the satellites move in orbits of low inclination with respect to Saturn's equatorial plane. Indeed, the orbital inclination of Phoebe exceeds 90° (Table 10.1), which means that its orbital motion is retrograde. It seems likely that Phoebe is a captured body, whereas the remainder formed as part of the Saturnian system. Except for Titan the satellites are fairly small, though a number are of a size between that of the Galileans and the small satellites of Jupiter. The satellites are further discussed in section 10.4.

Saturn, like Jupiter, possesses a substantial magnetosphere. A typical sunward position of the magnetopause in the plane of Saturn's equator is shown in Figure 10.1.

10.1 The magnetosphere of Saturn

Figure 10.2 shows the main features of the magnetosphere of Saturn. It is a good deal less extensive than the Jovian magnetosphere, because though the

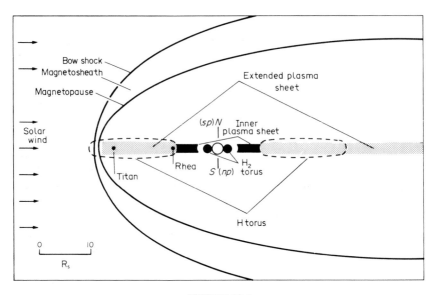

FIGURE 10.2
The Saturnian magnetosphere. The viewpoint is as in Figure 10.1. Magnetic
field lines are *not* shown.

number density of ions and electrons in the solar wind decrease with
increasing distance from the Sun, the magnetic field of Saturn is weaker than
that of Jupiter.

The magnetic dipole moment of Saturn is about 35 times less than that of
Jupiter. Per unit volume of planet the reduction factor is about 20, which, as
Saturn rotates about as fast as Jupiter, indicates that the fraction of Saturn's
volume that consists of *metallic* hydrogen is appreciably less than in the case
of Jupiter.

The polarity of the magnetic dipole field of Saturn is the same as that of
Jupiter, but the magnetic axis of Saturn is inclined with respect to the spin
axis by *less* than 1°. Also, the centre of the field is close to the centre of mass,
being displaced by only about 0.04 equatorial radii of Saturn (R_S), mainly
along the spin axis in the direction of the north pole.

The small offset could be the result of the dipole source being rather small
in volume and confined near to the centre of the planet. This is also indicated
by the spatial form of the field external to Saturn, which would be slightly
different were the dipole source at no great depth. This conclusion is clearly
consistent with the observed relatively small magnetic dipole moment.

The field external to Saturn is highly symmetrical around the magnetic
axis, more so than in the cases of the Earth, Mercury and Jupiter. The
dynamo theory predicts that the *dipole source* should not be particularly

symmetrical, and though at a sufficiently large distance the *field* will be symmetrical the possibility arises that in Saturn the dipole source itself is highly symmetrical. In this case, the dynamo theory predicts that the field will be decaying. One way around this difficulty is to assume that the field is at present reversing polarity. This has happened several times on the Earth, and may be accomplished by the field reducing to zero and then growing with the opposite polarity. Such a reversal might also account for the present close alignment between the magnetic and spin axes, which is otherwise hard to understand in terms of the dynamo theory.

The Saturnian magnetosphere contains neutral gas and plasma. It also contains highly energetic electrically charged dust in the ring plane, the source of which is unknown. Otherwise, all the same kinds of sources and losses and the same kinds of accelerating and decelerating mechanisms exist as in the Jovian magnetosphere, though there are significant differences of detail. For example, the number densities of ions and electrons are lower in the Saturnian magnetosphere, largely because of losses to the extensive ring system.

The neutral gas and plasma lie largely in the sheets and toruses shown in Figure 10.2 though it is important to realize that the boundaries are not as sharp as indicated, and the particle densities are not the same at all points within a given sheet or torus. All the sheets and toruses shown encircle Saturn.

The huge H torus extending inwards from beyond Titan to Rhea consists largely of neutral atoms of hydrogen, most of which are thought to have come from the atmosphere of Titan. Titan is probably also the source of nearly all the atoms and ions of *nitrogen* that are observed in the magnetosphere.

The other predominantly neutral torus, the H_2 torus in Figure 10.2, forms a tenuous atmosphere above and below the rings. It consists mainly of neutral hydrogen molecules (H_2) which are probably derived from water ice on the ring particles, photodissociated by solar uv radiation.

The two plasma sheets in Figure 10.2 are the result of leakage from the radiation belt (not shown) in much the same manner as in the Jovian magnetosphere. Beyond about $7R_S$ the magnetic field associated with these plasma sheets becomes significant compared to the dipole field of Saturn, but for the sake of clarity magnetic field lines are not shown in Figure 10.2.

Both Voyager spacecraft (Table 10.2) detected several sorts of previously unknown radio wave emission from the Saturnian system. In one sort it is possible to discern a sidereal periodic variation of 10h 39.4m, which is sufficiently close to the sidereal axial period of high latitude cloud features on Saturn to suggest that it is the axial period of the interior of Saturn. This is a similar case to that of Jupiter, and as in the Jovian case this is called the *System III sidereal axial period*. The source of these radio waves seems to be relatively small areas near the poles where aurora are particularly strong, though the generating mechanism of the waves and the mechanism of influence of the dipole source as implied by the periodic variations are not properly understood.

10.2 Saturn's clouds and atmosphere

There are many similarities between the clouds and atmosphere of Saturn and Jupiter, and therefore I shall concentrate here on the main differences between the two planets.

Figure 10.3 is a typical spacecraft picture of Saturn, and the broad similarity with Jupiter (Figure 9.3) is at once apparent. However, the cloud features on Saturn are rather less distinct, and this has made it difficult to obtain accurate rotation periods. Earth-based observations have yielded a prograde sidereal axial period of 10h 14.0m for equatorial regions (System I) and about 11h for polar regions (System II).

It is clear that, as on Jupiter, east–west winds are responsible for the difference between these periods. Figure 10.4 shows these winds, measured by Voyager 1 with respect to a cloud-top surface rotating at a speed corresponding to the System III period (section 10.1). The winds blow stronger than on Jupiter, are more predominantly from the west, and are rather more weakly correlated with the belts and zones. Also, the belts and

FIGURE 10.3
Saturn imaged by Voyager 1 from a range of 34×10^6 km. The rings appear elliptical because of the oblique view. North is to top left.

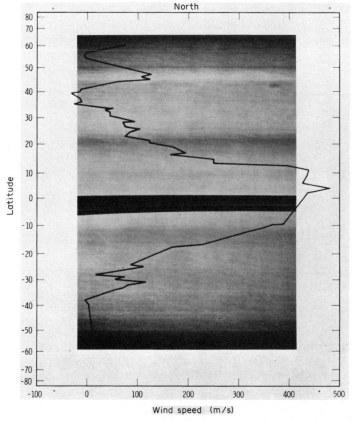

FIGURE 10.4

East–west wind speeds on Saturn based on Voyager 1 data. Speeds greater than zero correspond to flow from the west to the east.

zones extend to about 75° north and south of the equator, whereas on Jupiter they only extend to about 60°. Clearly there are some differences in the atmospheric circulation in the two planets though observations suggest that the basic circulation mechanisms are probably much the same. Possible reasons for the differences are the greater predominance of internal heat flow in Saturn, which radiates to space 2 to 3 times more energy than it receives from the Sun, the lower gravity, which could favour large-scale internal circulation, and a thicker molecular hydrogen "mantle", which would influence those upper atmospheric motions which are deep-seated. (A thicker "mantle" is consistent with the evidence outlined in section 10.1 that in Saturn the metallic hydrogen lies deeper than in Jupiter.) However, as yet there is no detailed understanding of these various possibilities.

The composition and vertical structure of the atmosphere of Saturn (Figure 10.5) are much the same as for Jupiter, but with a few interesting

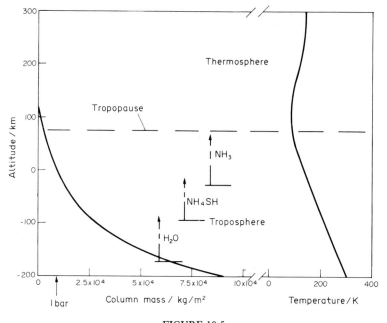

FIGURE 10.5
The vertical structure of the Saturnian atmosphere.

differences. First, the relative numbers of molecules in the atmosphere above the Saturnian clouds is as follows: 94% H_2; 6% He; 0.05% CH_4; 0.02% NH_3; plus traces of a few other gases. These figures show that helium is significantly less abundant than the 10% observed in the Jovian atmosphere, and I shall return to this point in section 10.3. Second, comparison of Figure 10.5 and 9.5 shows that the cloud bases lie deeper in the troposphere on Saturn than on Jupiter, and that on Saturn there is a greater mass of atmosphere between these bases and the tropopause. This is largely a consequence of the lower temperatures in Saturn's atmosphere, because of Saturn's greater distance from the Sun, with a tropopause at about the same pressure as in the Jovian atmosphere.

The relatively deep cloud bases on Saturn have led to one possible explanation of why the cloud features there are rather indistinct, as follows. The cloud features are associated with cloud *tops*, and these too will lie deep, because the main bulk of the clouds will extend a relatively small distance above their bases. Therefore, there will be a large thickness of atmosphere between the cloud tops and the tropopause, and into this thickness convection could carry the smaller cloud particles to yield a thin haze of considerable depth which would obscure cloud-top features.

Another explanation depends on the comparatively large mass of atmosphere above the cloud tops. A haze could be produced in this region through chemical reactions aided by solar uv radiation, the large mass of atmosphere providing large quantities of material for such chemistry to manipulate.

Although observations suggest that a haze, however produced, does indeed exist above the cloud tops, it is just possible that no such haze exists, in which case the explanation of the indistinctness of the cloud features could be that the production rate of coloured substances in the atmosphere is rather low, and therefore these substances are dissipated by atmospheric winds before they can reach sufficient concentrations in their production regions to yield bold markings. A lower production rate than on Jupiter could arise from the lower intensity of solar uv at Saturn and from its lower atmospheric temperatures, both of which result from Saturn's greater distance from the Sun.

In the period between the fly-bys of Voyagers 1 and 2 (Table 10.2) the cloud features in the northern hemisphere grew rather more distinct. This may be a seasonal change, the axial inclination of Saturn being 27°. However, no other appreciable changes were observed, indicating the deep-seatedness of many of the atmospheric phenomena.

10.3 The interior of Saturn

The mass of Saturn was first determined from the orbits of its satellites and later from its gravitational influence on the paths of fly-by spacecraft. The mass is $95.147M_E$. The equatorial radius at the 1 bar pressure level is 60 000 km and the polar radius at this pressure is 54 700 km making Saturn the most flattened of the planets. It actually spins slightly more *slowly* than Jupiter, and so the greater flattening must result from a difference in internal structure. A possible difference will be revealed shortly.

The mean density of Saturn is 705 kg/m³, and indicates that Saturn must consist largely of hydrogen because no other substance could yield such a low density in such a massive body. Jupiter's mean density is 1340 kg/m³, and therefore you might be tempted to think that hydrogen dominates Saturn even more than it dominates Jupiter. However, this is not the case: Saturn's mass is only 29.9% of that of Jupiter, and therefore the compression of the interior is considerably less.

Models of Saturn's interior have been constructed along similar lines to those of Jupiter. Figure 10.6 shows a typical example. You can see that its broad features are similar to those of the Jovian model in Figure 9.10. And like Jupiter, Saturn has no surface, solid or liquid, beneath the atmosphere.

Much the same sort of variations on this model are possible as in the Jovian case. There are, however, a few important differences between the Jovian and Saturnian models, as follows.

Because of its lower mass the pressures inside Saturn are lower than in Jupiter and therefore the transition to metallic hydrogen occurs at a greater

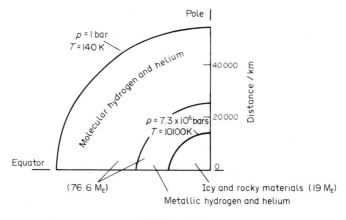

FIGURE 10.6
A model of Saturn.

depth in Saturn. This is consistent with the magnetic field data which suggest that the magnetic dipole source lies rather deep in Saturn and is rather small. Correspondingly there is a mantle of molecular hydrogen of greater thickness in Saturn. This leads to rather greater uncertainty in modelling Saturn because the equation of state of molecular hydrogen is more poorly known than that of metallic hydrogen.

It is possible that substances other than hydrogen and helium account for a larger fraction of Saturn's mass than such substances do in the case of Jupiter. The greater predominance of such substances adds to the uncertainty of modelling because the relative abundances of the various ingredients, their distribution within the inner regions of Saturn, and their equations of state, are poorly known.

The evolutionary sequence that yields the model of Saturn today in Figure 10.6 starts 4600 Ma ago with Saturn very hot from accretion, but because of Saturn's greater ratio of surface area to mass than that of Jupiter, Saturn cools sufficiently by about 2000 Ma ago for helium separation to begin. This releases heat and therefore slows down the rate of cooling of the interior. At a level of detail not shown in Figure 10.6 the helium abundance in the atmosphere in the region among and above the clouds has fallen to about half its initial value, with a progressively smaller depletion at greater depths. It is possible that other evolutionary sequences can lead today to the model in Figure 10.6, but the case for helium separation is supported by several strands of evidence.

First, the excess radiation from Saturn shows that there is a powerful energy source in Saturn, possibly greater per unit mass than in the case of Jupiter. The internal temperatures consistent with this, shown in outline in Figure 10.6, are much higher than would be expected if the heat of accretion alone is considered. The higher ratio of surface area to mass than for Jupiter means that Saturn cools quicker, and regardless of how hot Saturn became on accretion it would have cooled to yield its present heat losses by no later than about 2500 Ma ago. Solar choking of heat flow from the interior, outlined in section 9.3.3, can extend the cooling time but not by a large amount. Thus if Saturn is 4600 Ma old, as seems very likely from ages determined elsewhere in the Solar System, then an additional source of energy is required.

Second, the measured abundance of helium above the clouds is about half the solar relative abundance of He/H_2. The model predicts that helium separation accounts for this depletion, thus allowing Saturn as a whole to have a value of He/H_2 equal to the solar relative abundance. This very reasonable feature is built into the model illustrated by Figure 10.6. Without such separation Saturn as a whole would be depleted in helium, unlike Jupiter, and this would be hard to understand.

Third, the values of J_2 and the large flattening of Saturn can be obtained from models in which Saturn is depleted in helium by a factor of 2 *throughout* only by making rather unlikely assumptions about the interior. By contrast, models in which the depletion is largely confined to the outer regions can yield J_2 and f without making unlikely assumptions.

If, in spite of this strong evidence, helium separation has *not* occurred in Saturn, then the excess radiation can be explained by the gradual *solidification* of more and more metallic hydrogen near Saturn's core as Saturn cools, which would release latent heat and extend the cooling time. This possibility is denied to Jupiter because of its greater temperatures in these deep regions. This source of heat in Saturn may in any case be additional to that provided by the gravitational separation of helium.

10.4 The satellites of Saturn

I shall be mainly concerned with Titan which is of terrestrial planet size (Figure 1.2) and is the only satellite known to have a substantial atmosphere. The other satellites are *far* smaller.

10.4.1 Titan

The mass and radius of Titan have been measured, and the mean density is 1900 kg/m³. The only plausible major constituents are rocky and icy materials, and the mean density suggests roughly equal proportions. Of the common ices water is the most abundant and the least volatile and is therefore the most certain icy ingredient, though methane and ammonia may also be present. Carbon dioxide, which is also solid at temperatures typical of Titan, is unlikely to be present for reasons to be outlined in section 10.4.3.

Figure 10.7 shows a model of Titan, in which the rock has the same uncompressed density as the estimated uncompressed density of Io, and the ice is predominantly water. If Titan accreted homogeneously then the only plausible source of heat that could have caused differentiation is the heat of accretion. It is also possible that Titan accreted heterogeneously, or that it accreted homogeneously and remains in that state today, the model in Figure 10.7 then not applying.

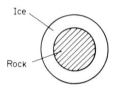

FIGURE 10.7
A model of Titan.

Titan has no magnetic field.

Observations have established that Titan has an atmosphere and that the atmosphere contains haze, possibly with a layer of cloud beneath. Observations have also established the atmospheric composition above the haze. The relative numbers of molecules is as follows: about 90% N_2; about 10% Ar; about 1% CH_4. Traces of other gases have been detected, and in many cases could have been produced from nitrogen and methane through chemical reactions aided by solar uv radiation. Neither water, nor ammonia, nor CO_2 have been detected, and in all cases this could result from the low partial pressures expected of these substances at the low surface temperatures of Titan, about 90 K. The partial pressure of methane at 90 K is much higher.

Some CH_4 will be photodissociated by solar uv radiation and the hydrogen liberated will escape to space. This could account for the neutral hydrogen torus within the orbit of Titan (section 10.1). Nitrogen can also escape, and as on Mars this happens with the aid of chemical reactions (section 3.4.2). This nitrogen probably accounts for much of the nitrogen in the magnetosphere. Loss of gases is enhanced by sputtering.

Figure 10.8 shows the atmospheric profile of Titan, determined largely by the Voyager spacecraft. The surface pressure is about 1.6 bar, but because the surface gravity on Titan is only about $\frac{1}{7}$ of that on the Earth the mass of the atmosphere per unit area of surface is about *10* times *greater* than on the Earth, where the surface pressure is about 1.0 bar. This massive atmosphere results in surface temperatures that vary little with latitude. The surface temperatures are slightly raised by a greenhouse effect due to nitrogen, hydrogen and methane.

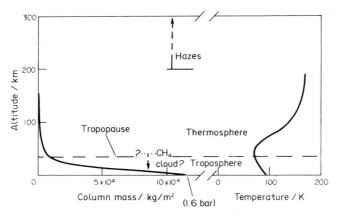

FIGURE 10.8
The vertical structure of Titan's atmosphere.

The haze is orange tinted and is probably a product of chemical reactions aided by solar uv radiation, though its composition is unknown. The gradual settling of haze particles to the surface and their continuous replenishment could have resulted over 4600 Ma in a layer of hydrocarbons several hundred metres deep at the surface. Methane may be able to condense in the atmosphere, to form a layer of cloud beneath the haze: it depends on its partial pressure and on the exact temperature profile, neither of which are known accurately enough to be certain whether methane clouds form. Such a cloud layer would reduce the methane content in the atmosphere above it. Thus, beneath the clouds methane could be considerably more abundant than the 1% observed above the haze. If it reaches about 8% at the surface then there could be solid or liquid methane deposits on the surface, the phase depending on the surface temperatures. If any such surface deposits exist then the surface temperature will determine the methane abundance in the lower atmosphere.

Some very subdued features have been seen in the haze, and there also seem to be changes in response to the seasons and in response to varying conditions in the magnetosphere.

Titan could have acquired its atmosphere as follows.

The argon could largely be the result of the radioactive decay of potassium in Titan's rocky component. The nitrogen could be the accumulated result of 4600 Ma of photodissociation of ammonia (NH_3) by solar uv radiation, the hydrogen escaping to space, the escape rate of nitrogen being far slower. A large fraction, perhaps all of Titan's supply of ammonia could have been lost in this way, though ammonia could still exist in solid form, still gradually feeding the atmosphere with nitrogen via photodissociation. The methane is the unmodified product of its own solid, more abundant in the atmosphere than the other common ices because of its greater volatility.

There is evidence that the temperatures on Titan have not been sufficiently high for water to have been an abundant atmospheric constituent. Had water been an abundant atmospheric constituent then the photodissociation of water by solar uv radiation would have yielded a slow input of oxygen to the atmosphere. This would have reacted with methane to produce water and carbon dioxide, and with nitrogen to produce nitrates and nitrites, yielding an atmosphere of far less mass than that observed and one dominated by carbon dioxide and water. However, you shall see in Chapter 15 that as the Sun evolves in the future its luminosity will increase. Therefore, temperatures on Titan will rise and the atmosphere may be transformed to CO_2 and water. Moreover, Titan may become warm enough for surface water to become liquid, in which case, given the rich variety of carbon compounds that probably exist on Titan's surface, it could then resemble the early Earth. Titan almost certainly has no biosphere today. But in this distant future . . . ?

10.4.2 The other satellites of Saturn

Some of the other satellites are comparable in size to the small Jovian satellites, and others are of sizes between these small satellites and the planetary-sized Galileans and Titan (Table 10.1). Some of these other satellites of Saturn have had their densities determined (Table 10.1) and they are less dense than Titan and are thus even more dominated by ices, particularly water.

Most of the named satellites have been imaged. Impact craters are common, and on some, particularly Enceladus, there is evidence of surface reworking. The source of energy for surface reworking is probably tidal. In the case of Enceladus the main tidal effect may be from the satellite Dione rather than Saturn, because Enceladus and Dione are in *orbital resonance*, a term to be outlined in section 11.5, where you will also learn that some of Saturn's satellites have important interactions with Saturn's rings.

Some of the satellites display hemispherical differences in colour and albedo, particularly Iapetus on which the hemisphere that faces in the direction of its orbital motion has an albedo of 4% to 5% and is tinted red, and the other hemisphere is less red and has an albedo of about 50%. It is not certain whether this difference is of internal origin or whether the forward face has preferentially collected material from the magnetosphere possibly originating from Phoebe, which also has a low albedo and a red tint.

10.4.3 The origin of the satellites of Saturn

Phoebe is separated by a considerable gulf from the other satellites (Figure

10.1). Its gravitational binding to Saturn is therefore not as tight as for the others and moreover its orbit is retrograde, eccentric and highly inclined. This strongly suggests that Phoebe was captured by Saturn, perhaps with the aid of gases around Saturn as outlined in section 9.4.4. Phoebe has a red tint, an albedo of about 5%, and is a good deal more spherical than many other bodies of its size (about 100 km radius). This suggests that Phoebe may be a comet remnant rather than an asteroid. Comets are a subject for Chapter 14.

It would be very difficult to capture the other satellites into orbits so close, and thus so tightly bound to Saturn. This at once suggests that they formed as part of the Saturnian system. The distribution of sizes among this group of Saturn's satellites is substantially different from that of the corresponding, inner group of Jupiter's satellites: compare Tables 9.1 and 10.1. A plausible explanation for this difference is based on the likelihood that because of its smaller mass Saturn formed slower than Jupiter. The details will not concern us.

The smaller mass of Saturn can also explain why ices are abundant in the Saturnian satellites from Rhea inwards, whereas at comparable distances from Jupiter, from Europa inwards, ices are far less abundant. The smaller mass of Saturn means *either* that it was less luminous just after accretion than was Jupiter and therefore it was cool enough for ices to condense on these inner satellites, *or* that if the luminosity of these giant planets subsequently grew to give a brief peak then Saturn never became as luminous as Jupiter and therefore only Jupiter drove most of the ices from its inner satellites.

The retention of a good deal of water by Ganymede and Callisto is readily understood as a consequence of their greater distances from Jupiter, and it is to be expected that methane and ammonia, even if they survived Jovian heat, would have been driven off by the heat of the Sun because of their much greater volatility. However, today the temperatures are too low even for water to yield an appreciable atmospheric pressure: this is also the case for Titan. Because of its greater distance from the Sun and the lower luminosity of Saturn Titan retained the more volatile ices methane and ammonia and you have seen that it is likely that the present atmosphere of Titan is largely derived from these.

The absence of argon on Ganymede and Callisto is not understood.

The ice/rock ratio on Titan is less than would be expected from abundances based on the relative solar abundances of the elements that make up the ices and the rocks, indicating that only some of the available methane and ammonia is present today.

Carbon dioxide does not feature here because during the formation of the Jovian and Saturnian systems it is clear that hydrogen was abundant, and in such an environment carbon is mostly bound in CH_4 and oxygen mostly in H_2O.

10.5 Impact cratering in the outer Solar System

The craters on the satellites of Jupiter and Saturn provide the beginnings of an observational basis on which cratering can be considered throughout a much larger part of the Solar System than the terrestrial region alone.

In Chapter 8 you saw that the present cratering of the terrestrial planets is due to asteroids and to comet debris. By contrast, calculations suggest that in the outer Solar System cratering is largely the result of comet debris and of comets in the pre-debris stage, with asteroids accounting for no more than about 1% of impacts in the vicinity of Jupiter and Saturn, and an even smaller fraction further out.

The *exposure* of a planetary satellite to cratering depends on the position of its parent planet in the Solar System and, particularly in the case of the giant and subgiant planets, on its distance from its present planet, because a massive parent planet can strongly influence the orbits of impacting bodies. The *cratering rate*, as outlined in Chapter 8, depends also on the properties of the satellite. It has been estimated that the cratering rate on the Jovian and Saturnian satellites, in the units of Figure 8.2, varies from a high of about 5 on Io, to about 0.3 on Rhea, to even less on Iapetus. In Figure 8.2 you can see that the corresponding rates on the terrestrial planets lie between 1.9 and 2.9.

Many of the satellite surfaces have been reworked: for example, Io has no known impact craters at all in spite of the estimated high rate of crater formation. Other surfaces display an intermediate number density of craters and in some cases, assuming that the present cratering rates have always been the same, the surfaces must have been accumulating craters for several 1000 Ma. Other surfaces are so heavily cratered that, at present cratering rates, they would have needed to accumulate craters for well over 4600 Ma: clearly, there is evidence here for a higher rate of cratering in the past.

This raises the question of whether any late heavy bombardment that *might* have pervaded the inner Solar System also pervaded the outer Solar System as well. Unfortunately the observational evidence is too scanty and there are too many difficulties of interpretation for an answer based on observations to be given at present. A separate line of investigation is theoretical, and this line is intimately entwined with theories of the origin of the Solar System, the subject of Chapter 15.

10.6 Summary

Saturn is broadly similar to Jupiter. It is a rapidly rotating massive planet consisting largely of hydrogen, and very probably has a hot interior. One of the main differences between Jupiter and Saturn is that Saturn's magnetic field seems to be deeper seated and probably originates in a very symmetrical dipole source. Another is that cloud features on Saturn are more subdued than on Jupiter. It is also likely that metallic hydrogen accounts for a smaller fraction of Saturn's volume than in the case of Jupiter, and that downward separation of helium in Saturn has been a significant source of interior heat. The differences between these two planets arise from Saturn's smaller mass and from its greater distance from the Sun. These two factors probably also explain why Titan has an atmosphere but the Galileans do not, and may also account for other differences between the non-captured satellites of the two systems.

10.7 Questions

1. List the similarities and differences between Jupiter and Saturn.
2. List the similarities and differences between the satellite systems of Jupiter and Saturn.
3. What possible reasons are there for the cloud features on Saturn being more subdued than on Jupiter?
4. Outline the evidence that helium separation has occurred in Saturn.
5. What are the likely reasons for there being nitrogen in Titan's atmosphere, but negligible amounts of carbon dioxide and water?

11

PLANETARY RINGS

11.1 The ring systems of Jupiter, Saturn and Uranus

You have already learned that Jupiter and Saturn have rings. The only other planet known to have rings is Uranus, five of which were discovered by accident in 1977 during Earth-based observations of a stellar occultation by Uranus, a further four being added during observations of subsequent occultations.

Figure 11.1 shows the main features of the three ring systems, except that I have had to omit a very faint and broad outer ring of Jupiter and the faint and broad outer E ring of Saturn. All three ring systems lie close to the equatorial planes of their respective planets, and compared to the ring systems' diameters they are extremely thin, with typical thickness of only about a kilometre.

Detailed analyses of the solar radiation scattered by the rings of Jupiter and Saturn have established that the Jovian rings probably consist mainly of

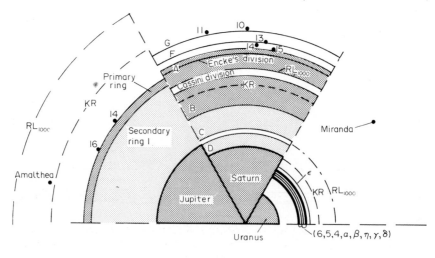

FIGURE 11.1
Ring systems.

micron-sized particles and that the rings of Saturn consist mainly of larger particles with radii in the range 5 mm to 5 m. The radii of the Uranian ring particles are unknown, but it is presumed that they are no more than a few metres radius. Similar analyses have also revealed that the ring particles, or at least their surfaces, consist largely of rocky or carbonaceous material in the Jovian case and largely of fairly clean water ice in the case of Saturn. The rings of Uranus are extremely dark, the average ring albedo being no greater than about 2.5%, which roughly corresponds to matt black paint. Carbonaceous materials could account for such a low albedo.

Figure 11.1 does no justice to the intricacy of the ring systems, particularly in the case of Saturn. The single ring discovered by Huyghens in 1659 corresponds to rings A and B, and you can see from Figure 11.1 that several other rings have now been added. However, it is now known, largely from the two Voyager missions, that each of these rings consists of numerous narrow, concentric *ringlets* and that the major divisions between the rings also contain a few of such ringlets. Several thousand ringlets are known, and it is clear that beyond the present resolution limit of about 100 metres yet finer structure may lurk. Several ringlets are slightly elliptical, others vary in brightness from point to point, and a few, notably those that make up the narrow F ring, are kinked, or wavy. In many parts of the ring system there are very narrow gaps between adjacent ringlets or groups of ringlets, and these are called *gaplets*. Figure 11.2 gives some impression of this intricacy. The *inset* shows a set of diffuse radial markings called spokes, any particular spoke being a transitory feature. These spokes are thought to result from interactions between the ring particles, the magnetospheric ions and electrons, and Saturn's magnetic dipole field.

The Jovian ring system has not yet revealed as many intricacies, though it has been less thoroughly studied and is much more tenuous than the Saturnian system.

The nine rings of Uranus are remarkable for their narrowness, ranging from no more than about 3 km wide to about 100 km wide, the widest being the ε (epsilon) ring (Figure 11.1). At least six of the rings are elliptical rather than circular, and they also vary in width around their circumferences. At least three of the rings have internal structure, possibly corresponding to ringlets.

Most of the intricate structural features of the ring systems are thought to result from a variety of gravitational interactions. Some of the relevant interactions are between the ring particles whilst others also involve the satellites of the planet.

Satellites can also be important *sources* of ring particles, particularly Io in the Jovian case and Enceladus in the case of Saturn. Such sources help to offset the loss of ring particles, which undoubtedly occurs by a variety of means.

FIGURE 11.2
The rings of Saturn imaged by Voyager 1 from a range of 1.5×10^6 km. The
rings appear elliptical because of the oblique view. North is to top right. Inset:
ring "spokes", imaged by Voyager 1 from a range of 24×10^6 km.

11.2 The origin of ring systems

In section 6.1.1 you saw that if a satellite is orbiting closer to a planet than
the orbit of Keplerian co-rotation then the satellite will gradually spiral
towards the planet. Ultimately it will pass within the Roche-limit for a body
of its density, and if it is not too small it will then be tidally disrupted. The
fragments will tend to collide, producing further fragmentation, the J_2 term
in the planet's gravitational field will rapidly concentrate the particles into the
equatorial plane of the planet, and subsequent interactions between these
particles will cause the ring to spread out. A variety of mechanisms rid the
ring of very large and very small particles, and the result, allowing for the
later development of intricate structural features, is ring systems of the sorts
that we observe.

The Roche-limit is clearly of central importance to this source of ring
particles. In Figure 11.1 the Roche limit for icy bodies is shown for each
planet (RL_{1000}) and also the orbit of Keplerian co-rotation (KR). In the case

of Saturn there is an appreciable amount of ring material outside of RL_{1000}, but this can be explained by the outward spreading of the ring particles because of a copious quantity initially within RL_{1000}. For Jupiter and Uranus the Roche-limit for rocky bodies is the more appropriate, and in both cases this lies just outside the ring systems shown.

There are, however, two other ways of producing ring particles. First, they can be left over from planetary accretion. After a planet has formed it is very likely that debris will be left in orbit around it. Any such debris which lies within the Roche-limit, or not too far outside it, will fail to accrete into larger bodies. Second, and again after a planet has formed, the ring particles could be captured one by one or in groups. Those which found their way near to or within the Roche-limit would, as before, fail to accrete into larger bodies.

The Roche-limit seems to be of little relevance to the far-flung ring of Jupiter which extends outwards from about $3.5R_J$. The origin of this ring, which in any case is very tenuous, is obscure.

The *differences* between the three-ring systems slightly favours the accretional left-over mechanism for producing ring particles, as follows.

In the period following planetary formation it is likely that Jupiter was far more luminous than Saturn. Therefore the icy materials close to Jupiter would have been kept in the gaseous phase until they had largely dissipated, whereas those around Saturn would have condensed to produce the comparatively massive system of icy rings which we observe today. Jupiter would only have the less abundant rocky materials.

Uranus should also have been surrounded by abundant icy material, and like Saturn it should have been cool enough for it to condense. However, the condensation of icy materials is greatly speeded by the existence of condensation nuclei, and these could be provided by non-icy condensates. However Uranus, because its mass is considerably less than that of Saturn, may never have had around it more than a trace of non-icy materials. Therefore most of the icy material could have been dissipated before condensation could occur, leaving Uranus with the non-icy low mass ring system that we see today.

Neptune could have a ring system comparable to that of Uranus, though the remoteness of Neptune makes it difficult to establish whether any such system exists.

The terrestrial planets seem to be devoid of any rings. Perhaps this is because their low masses resulted in very sparse rocky left-overs, and because they are too near the Sun for icy materials to have condensed on any such remnants.

11.3 Summary

Planetary rings could be produced in a variety of ways, the tidal instability of large bodies inside the Roche-limit playing a role in all present theories. The more massive the planet, and the cooler the conditions in its vicinity, the more extensive the ring system is likely to be. Most of the intricate structural features in the rings are probably the result of a variety of gravitational interactions.

11.4 Questions

1. Outline the various ways in which ring systems can originate.
2. What are the main structural features of the ring system of Jupiter, Saturn and Uranus?
3. Explain why the albedo of Jupiter's rings is lower than that of Saturn's rings.

12

THE SYSTEM OF URANUS

"On Tuesday the 13th of March, between ten and eleven in the evening, while I was examining the small stars in the neighbourhood of H Geminorum, I perceived one that appeared visibly larger than the rest: being struck with its uncommon magnitude, I compared it to H Geminorum and the small star in the quartile between Auriga and Gemini, and finding it so much larger than either of them, suspected it to be a comet."

This is the opening of a scientific paper written by the Germano-British astronomer William Herschel (1738–1822, father of John Herschel). It was read to the Royal Society of London by the British physician William Watson (1744–1825) on 26 April 1781. Thus was announced to the world the discovery, not of a comet, but what was soon shown to be a planet. Its true nature was clear by May 1781 after its orbit and large size had been established. The name "Uranus", in mythology the father of Saturn and first ruler of Olympus, was first suggested by Bode (of Titius-Bode law) in 1784, but the name was not everywhere accepted for several decades.

Uranus is the first planet to be discovered in recorded history, and though it had been recorded as a star on several earlier occasions it is not hard to understand why it remained unrecognized for so long: it is twice as far as Saturn from the Sun and is only 43% of Saturn's radius making it just about visible to the best unaided eyes in good seeing conditions, though because of its great distance from the Sun it takes over 84 years to complete an orbit and therefore moves extremely slowly against the stellar background. It is therefore easily mistaken for a faint star.

But Uranus is readily seen in a small telescope, and under high magnification can be distinguished from the stars because Uranus shows a disc which increases with magnification, whereas the stars remain as hard points of light. Herschel discovered Uranus with a reflecting telescope which he had made himself, in which the main mirror was only 157 mm diameter, a size comparable to the mirror sizes used by many amateur astronomers today. The discovery was made with an eyepiece that gave an angular magnification of 227. At the time, Herschel was making a systematic survey of the stars with particular emphasis on cataloguing *double stars*, that is, two stars which are *either* close to each other *or* which merely appear to be close together from our vantage-point, the one lying far beyond the other.

263

William Herschel's life is a remarkable and fascinating one during the course of which he made many contributions to astronomy of long-lasting importance. Though he was 42 when he discovered Uranus this came near the beginning of his professional career as an astronomer and marked the end of his earlier career as a musician. His major astronomical work related to stars and galaxies and to the construction of the finest telescopes of the day with which he became one of finest observers of his time. But he continued to make further contributions to planetary astronomy also, and in 1787 at his new observatory in Slough, England, he discovered two satellites of Uranus, subsequently named Titania and Oberon by Herschel's son John Herschel, who was also an accomplished astronomer and in addition made major contributions to the development of photography. John Herschel also named the next two Uranian satellites to be discovered, by the British astronomer William Lassell (1799–1880) at Liverpool, England, in 1851. He called them Ariel and Umbriel. These names are of fairies in works by Pope and Shakespeare. This naming tradition continued with Miranda, the most recent Uranian satellite to be discovered, by the Dutch-American astronomer Gerard Peter Kuiper (1905–1973) in 1948.

Table 12.1 lists the Uranian satellites. It is expected that tidal interactions with Uranus have given rise to synchronous rotation for all five. Their masses are unknown, but their mass *ratios* have been determined from small but measured gravitational effects they have on each other's orbits. Indeed, the accuracy of the mass ratios is limited not by the observations but by the theoretical understanding required to extract the ratios.

Infrared spectrometry from Earth has revealed that clean water frost lies on the surfaces of all five satellites. On Umbriel the spectral signatures are less sharp than on the others, suggesting that *either* the frost is mixed with other substances *or* it covers a smaller fraction of the surface. By assuming on the basis of such spectroscopic data reasonable values for the albedo of a satellite its radius can be calculated from its brightness, which depends on its albedo, its area and the intensity of solar radiation at Uranus's distance from the Sun. These calculated radii are shown in Table 12.1.

The satellites probably consist of roughly equal mixtures of icy and rocky materials, and therefore have densities between 1500 and 2000 kg/m^3. Their masses are certainly small enough to make it very unlikely that these satellites ever melted during accretion or became hot enough to melt subsequently. They are unlikely to have any atmosphere.

You can see from Table 12.1 that the orbits of the satellites all lie in almost the same plane and this is illustrated in Figure 12.1 (a). But this plane is inclined at 97.92° with respect to the orbital plane of Uranus around the Sun. Therefore their orbits go through the cycle shown in Figure 12.1 (b). As viewed from the Earth the face on view next happens in 1987. The proximity of the satellites to Uranus suggests that they were not captured, but formed as

TABLE 12.1
The satellites of Uranus

No.	Name	Year of discovery	a/(1000 km)	e	i/°	Sidereal orbital period/days	Radius/km	Density/ (kg/m³)
5	Miranda	1948	130	0.017	3.4	1.414	160	?
1	Ariel	1851	192	0.0028	0.0	2.520	470	?
2	Umbriel	1851	267	0.0035	0.0	4.144	410	?
3	Titania	1787	438	0.0024	0.0	8.706	520	?
4	Oberon	1787	586	0.0007	0.0	13.46	460	?

Uranus, equatorial radius at 1 bar = 25 900 km.
ε ring, radius = 51 200 km.
Notes: a = semi-major axis of orbit; e = eccentricity of orbit;
i = inclination of orbit with respect to Uranus's equatorial plane;
? = not known.

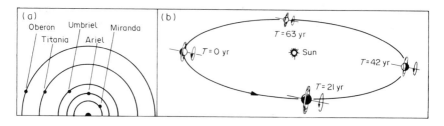

FIGURE 12.1
(a) The orbits of the satellites of Uranus. (b) The axial inclination of Uranus. Much of the eccentricity of Uranus' orbit and those of its satellites arises from the oblique view.

a part of the Uranian system. The rings of Uranus lie in the same plane as the satellite orbits.

In the case of Jupiter and Saturn, the rings and the inner satellites lie in the equatorial plane of the planet. Moreover, the J_2 component of the gravity field of a planet, which results from the planet's axial spin (Figure 3.5), is expected to rapidly place the orbits of nearby satellites and rings into the equatorial plane. Therefore, it seems very likely that the equatorial plane of Uranus lies in the plane of its satellites and rings, in which case the axial inclination of Uranus is about 98°, as shown in Figure 12.1 (b). This means that the spin of Uranus is retrograde.

This possibility was first suggested by the French scientist Pierre Simon Laplace (1749–1827) when only two satellites and no rings had been discovered.

Further support for this high axial inclination comes from the ε ring. This ring is elliptical and its *pericentre* (equivalent to the perihelion of a planet) is observed to precess in the plane of the ring rather like the orbit of Mercury (section 7.0) though in this case the cause is the J_2 component of the gravitational field of Uranus, and this precession can be shown to imply that the equatorial plane of Uranus lies in the plane of this precession. More support comes from spectroscopic studies of Uranus, to be outlined in section 12.2, which show that the spin axis lies within 4° of the 98° value. The most obvious possible source of support, the observation of surface or cloud features, provides no conclusive evidence, the markings being too indistinct.

It is not known for certain how Uranus came to have an axial inclination of 98°. If it originally had a far smaller inclination then it would require an appropriately oblique impact by a body about the mass of the Earth to produce the present inclination. Such an event is very unlikely now, but in the early days of the Solar System there was probably far more matter flying around than there is today, as is clear from impact cratering records, and therefore this is a plausible explanation.

In the following sections you will find that there are many serious gaps in our knowledge of Uranus. It is so remote that it is difficult to observe from the Earth. But in 1986 Voyager 2 will fly by Uranus, making its closest approach on 24 January, and at about the same time a large telescope, *Space Telescope*, should be ready in orbit around the Earth and should obtain pictures of Uranus comparable to those that Voyager is expected to yield.

12.1 The atmosphere of Uranus

It has long been known from a variety of observations that Uranus has a substantial atmosphere, and that, at least at solar wavelengths, we see only a haze, and perhaps cloud tops beneath it. The haze need not be dense, because the atmosphere is undoubtedly very deep (section 12.2), and though Uranus possesses a surface it would be hidden from us even if there are no clouds.

Spectrometric observations have shown that the atmosphere consists mainly of molecular hydrogen. Helium has not been observed directly, its spectral signatures are too weak, but its existence has been inferred from other spectrometric observations. Helium could account for anything between a few percent to about 20% of the number of hydrogen molecules, a range which includes the solar relative abundance of He/H_2 of about 11%. Methane has been observed, and in the atmosphere above any cloud is likely to account for nearly all the carbon present. The corresponding relative abundance of C/H_2 is somewhere in the range 0.13% to 0.40%. The solar relative abundances of C and H correspond to a C/H_2 ratio of about 0.07%, and so the atmosphere of Uranus above the clouds is *enriched in carbon*. In Jupiter and Saturn the C/H_2 ratios are close to the solar value.

Ammonia has not been detected spectrometrically, but its abundance has been inferred from radiometric data, and it is clear that if ammonia accounts for most of the nitrogen in Uranus's atmosphere then the whole atmosphere is *depleted in nitrogen*: the estimated N/H_2 ratio is roughly 100 times less than solar relative abundance of N/H_2.

The depletion of nitrogen and the enrichment of carbon means that the C/N ratio is several hundred times larger than the solar relative abundance of C/N. Moreover, whereas the nitrogen depletion applies to the whole atmosphere of Uranus, the carbon enrichment applies only to the atmosphere above any clouds and therefore if there are any clouds which largely consist of CH_4 then methane would be even more abundant in the atmosphere beneath the clouds and the C/N ratio would be even more different from the solar relative abundance.

It is expected that the C/N ratio in a planet should be fairly similar to the solar relative abundance of C/N. This is the case for Jupiter and Saturn, and it is also the case for Venus, the Earth and Mars, as Figure 5.2 can show. The value of C/N for Uranus is therefore problematical: it is not the high value of C/H_2 that poses the problem, because there are many ways in which a planet can be depleted in H_2; the problem is posed by the apparently low value of N/H_2. I shall return to this problem in section 12.2.3.

Figure 12.2 shows atmospheric profiles for Uranus consistent with the observational constraints. These constraints are rather weak and this has resulted in the range of temperature profiles shown and in the question marks attached to the haze altitude range and to the clouds, though clouds of some sort probably exist somewhere in the atmosphere. Methane does not condense in the atmospheres of Jupiter and Saturn because the temperatures are too high. The atmosphere of Uranus is considerably cooler and therefore condensation of methane to form cloud particles is possible, depending on its partial pressure. However, the atmosphere of Uranus is too poorly known for it to be certain whether any such clouds consist of methane. If they do, then the abundance of methane beneath the clouds will exceed the measured abundance above. The clouds could also consist of ammonia, or hydrogen sulphide. Ammonium hydrosulphide and water are ruled out because they are insufficiently volatile for their partial pressures at the low Uranium temperatures to be sufficiently large for condensation to occur at such comparatively high levels in the atmosphere. The presence of ammonia clouds would yield a greater ammonia abundance beneath them than above. However, this ammonia has already been included in the above estimate of atmospheric ammonia, and therefore any such clouds do not solve the problem of ammonia depletion.

The detached haze (Figure 12.2) is a fairly certain feature, and seems to consist of micron-sized particles. However, its composition is uncertain. Methane could possibly condense in this region of the atmosphere, or it may be a substance produced through chemical reactions aided by solar uv radiation. Detailed comparisons with Neptune suggest that there is a component in the atmospheric haze of Uranus that does not occur in the atmospheric haze of Neptune.

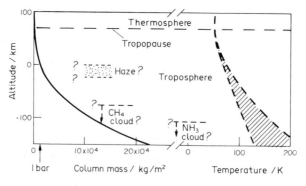

FIGURE 12.2
The vertical structure of the Uranian atmosphere.

Outside of the haze layer in Figure 12.2 there is very little haze. One consequence of this is that the green tint imparted by gaseous methane is readily visible in a telescope.

In Section 12.2 you shall see that in the troposphere of Uranus the temperature probably follows the adiabatic lapse rate, but it is not certain whether this persists into the relatively high altitude domain of Figure 12.2. The temperature at the tropopause is about 50 K, and though the temperatures in the thermosphere are uncertain it is clear that the thermosphere is surprisingly cool, indicating a substance that radiates efficiently to space at planetary wavelengths. Haze particles are poor radiators and therefore tend to *increase* thermosphere temperatures and thus a *lack* of haze in the thermosphere would help it to keep cool.

It is expected that, high in the thermosphere, there exists an ionosphere. At this distance from the Sun ion production by solar uv radiation is sufficiently slight that ion production by cosmic rays is significant in comparison.

Very few features have ever been seen in the Uranian haze/clouds. This is hardly surprising, because Uranus as seen from the Earth has about the same angular diameter as the Great Red Spot on Jupiter. Moreover, the haze itself will obscure any cloud features and the few features which have been seen probably correspond to non-uniformities within the haze. For example, the polar region currently presented to the Earth is brighter at solar wavelengths than the rest of the disc, and images at infrared wavelengths obtained in the early 1980s showed that this pole, which was then nearly pointing Sunwards, was surrounded by a slightly cooler collar. This could be the result of warmed atmosphere rising above the pole and descending through the surrounding region.

Seasonal variations on Uranus, because of its high axial inclination, are likely to be different from on any other planet, with correspondingly unique patterns of atmospheric circulation. A seasonal variation has been observed in the intensity of radiation at *planetary* wavelengths, which up to the early 1980s has consistently increased since 1966 when Uranus had its equator pointing towards the Sun. It can be shown that a possible explanation is a gradual increase in ammonia depletion, from a *slightly* less depleted level in 1966. However, it is not clear by what means such a change in depletion is caused. Alternatively, there could be a permanent dependence on latitude of the intensity of radiation emitted by Uranus at planetary wavelengths, and the seasonal change is then because of the changing aspect of Uranus as seen from the Earth. Such radiation largely originates deep in the atmosphere where the Sun cannot readily make seasonal changes and thus a permanent latitude dependence is plausible.

Since 1966 Uranus has also been getting brighter at *solar* wavelengths. This too could be the result of the changing aspect. Indeed this is likely because of the known polar brightness at such wavelengths and because of the flattening of Uranus.

Non-seasonal changes are also known. For example, the periodic variations in brightness at solar wavelengths, associated with the axial rotation of Uranus, are sometimes more marked than at others.

12.2 The interior of Uranus

12.2.1 Observational evidence

The mass of Uranus has long been known from the orbits of its satellites, and is 14.54 M_E. The polar and equatorial radii have been obtained from stellar occultations, and from images obtained by the telescope *Stratoscope II*, which was placed high in the Earth's atmosphere on a balloon in 1970. The measurements, adjusted to the 1 bar pressure level, yield about 25 900 km for the equatorial radius, and about 25 300 km for the polar radius. The uncertainties in these radii are such that the flattening could be anything from 0.01 to 0.04.

The mean density is 1250 kg/m^3. This rules out a predominantly rocky or iron planet. However, hydrogen cannot be as dominant as in Jupiter or Saturn because the mass of Uranus is considerably less and therefore hydrogen would not be compressed in the interior to a sufficiently high density to be able to account for most of the mass. Ices must account for most of Uranus's mass, with lesser quantities of rocky materials and hydrogen and helium. Thus, Uranus is considerably depleted in hydrogen and helium with respect to solar relative abundances.

A value of J_2 for Uranus was first obtained from the motions of the satellites Miranda Ariel and Umbriel. However, since the discovery of the rings a value of J_2 with a smaller observational uncertainty has been obtained from the ϵ ring. This ring is slightly elliptical, and the semi-major axis of this ellipse precesses in the manner outlined earlier, the departure of the gravity field of Uranus from spherical symmetry causing the precession of the ϵ ring.

The value of J_2 is smaller than for Jupiter and Saturn and it is the dominant coefficient. On its own it is not a very useful constraint on models of the interior, but if it can be used to calculate the polar moment of inertia C then a much more useful constraint will be obtained.

The only way to obtain C for Uranus would be to assume hydrostatic equilibrium, and accordingly calculate C from the mass and equatorial radius of Uranus, plus any two of the flattening f, the axial period T_a, and J_2, the value of J_2 being sufficiently well known to be one of the two.

The axial period of Uranus has not been obtained by observation of cloud features because no suitable features have been observable for long enough periods. Nor have any non-thermal radio waves, periodic or otherwise, been detected. Two methods have been used. First, in *radiometric methods* the intensity of the solar radiation scattered by Uranus is accurately recorded and then scrutinized for periodic variations. Almost any non-uniformity of albedo across the disc would give rise to such variations as Uranus rotates. The periodic variations are very slight and it is therefore difficult to determine the periods accurately. The variations can sometimes be enhanced if radiation over a narrow wavelength range is examined. Second, in *spectrometric methods* use is made of the Doppler-shifts of spectral lines in the radiation from Uranus. There is a variety of methods, some of which yield the orientation of the spin axis in addition to the spin rate. (You will appreciate that there are some parallels here with the radar methods outlined in section 4.1.1.)

The most recent measurements using *radiometric methods* have yielded values around 24 hours for the sidereal axial period of Uranus. Such methods are particularly prone to "contamination" by the Earth's axial period, and so these Earth-like values must be regarded with some reservation. Some of the recent measurements using *spectroscopic methods* have also yielded periods around 24 hours, but others have yielded periods around 16 hours. The reason for this difference is not known. The true value probably lies between 13 hours and 24 hours, but at present it cannot be defined more precisely. Moreover, if there is a persistent east–west wind then this range may not include the value appropriate to the interior.

This considerable uncertainty in T_a, and the comparable uncertainty in f, means that the assumption that Uranus is in hydrostatic equilibrium cannot be checked by observation, though it is certain that a planet as massive as Uranus will be in such a state. More seriously, the range of calculated value of C is *far* too wide to provide a useful constraint on model building and therefore J_2, f and T_a have to be used separately as rather weak constraints.

No radiation excess from Uranus has been observed, which places an upper limit of about 15% of the absorbed solar radiation on any such excess. This does *not* necessarily mean that the interior of Uranus is cool: flow of heat from the interior could be largely blocked by some means (section 12.2.2).

The only foreseeable way in which a Uranian magnetic field could be detected from the Earth would be via the detection of non-thermal radio waves, as was the case for Jupiter (section 9.1.2). No such waves have yet been detected but this places no useful upper limit on the internal magnetic dipole moment. In 1986 when Voyager 2 flies by Uranus the spin axis will be pointing into the solar wind and therefore if Uranus has a dipole field with a small inclination with respect to this axis then it should yield a magnetosphere of a kind not previously seen in the Solar System.

12.2.2 Models of the Uranian interior

Models of Uranus are constructed along broadly similar lines to those of

Jupiter and Saturn. As in the case of these giant planets most of the recent models have a liquid interior fully convective through most of its bulk and therefore with an adiabatic lapse-rate up to some level near the tropopause.

There are two justifications for such "hot" models. First, the fairly large mass of Uranus means that it probably became fairly hot during accretion or perhaps subsequently as a result of differentiation, or by both means, and it is plausible that it cooled sufficiently slowly to have a moderately hot and largely convective interior today. Second, Neptune, as you shall see in Chapter 13, is fairly similar to Uranus in mass, density and atmospheric composition, and yet Neptune clearly has a radiation excess which is strong evidence that it has a moderately hot convective interior. This suggests that Uranus should also have a moderately hot convective interior, and that the heat from the interior is to a large extent being blocked. Solar choking (section 9.3.3) is a plausible blocking mechanism.

Figure 12.3 is a "moderately hot" model of Uranus. This model is

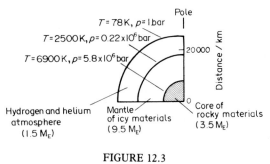

FIGURE 12.3
A model of Uranus.

consistent with the rather weak observational constraints, and also with evolution from a hot state immediately after accretion 4600 Ma ago, or from a hot state developing soon afterwards such as through differentiation. In either case the present cooling rate is significantly reduced by solar choking. Indeed, the solar choking may have been sufficiently effective to make the outer regions of Uranus non-convective in relatively recent times. In the case of Neptune it can be shown to be plausible that because of Neptune's greater distance from the Sun solar choking is less effective and thus we can "see" the moderately hot interior today.

Because of the weak observational constraints considerable departures from the model in Figure 12.3 are possible. Indeed, it is only on grounds of solar relative abundances that models of Uranus can be ruled out in which Uranus, below a fairly shallow atmosphere, consists largely of helium. A far more likely variant is that the icy materials are mixed with substantial amounts of hydrogen and helium, the predominantly hydrogen atmosphere in Figure 12.3 then being much less deep.

However, there are some broad features that any model should probably incorporate.

Thus, it is likely that the mass fraction of icy materials lies between 25% and 70% and that rocky materials constitute a mass fraction between 15% and 50%, the greater the rocky mass fraction the smaller the icy mass fraction. It is also likely that the rocky materials are largely concentrated in a central core.

In moderately hot models the core would probably be liquid, and any iron could be the source of a substantial magnetic dipole moment. Any such magnetic moment could be supplemented by water in the mantle near the core where temperatures may be sufficiently high to produce an appreciable degree of ionization of water molecules. This supplementation is also possible in Jupiter and Saturn. In some models of Uranus hydrogen exists at sufficient depth to raise the possibility that it is metallic because of the high pressures to which it is then exposed.

At the distance from the Sun at which Uranus formed it is likely that the temperatures were no more than a few hundred Kelvin and that the PFM was at a fairly low pressure. In these conditions, from a mixture of elements with solar relative abundances, if chemical equilbrium is achieved then the relative numbers of molecules in the PFM would have been as follows: 98.5% H_2 plus He; 0.6% H_2O; 0.4% CH_4, 0.4% rocky materials (including iron); 0.1% NH_3; plus traces of other substances. It is clear that Uranus is very depleted in hydrogen and helium, a feature of some importance to be taken up in Chapter 15.

The *overall* icy/rocky ratio in models of Uranus is roughly that expected in the PFM, but the question arises of whether ammonia is underabundant in comparison with the other abundant icy materials, that is, water and methane. You have seen that, in the *atmosphere*, the N/C ratio is considerably less than the solar relative abundance. Is this the case for the whole planet?

12.2.3 The ammonia problem

If Uranus is moderately hot and convective then it is difficult to see how there could be a substantial difference between the abundance of ammonia in the atmosphere and in the mantle of icy materials. One possibility is that ammonia has been removed from the atmosphere by hydrogen sulphide (H_2S), which as you saw in relation to the Jovian clouds reacts with ammonia to produce ammonium hydrosulphide (NH_4SH). This would condense in the Uranian atmosphere and would precipitate downwards. The quantity of H_2S required corresponds to an S/N ratio roughly 7 times the solar relative abundance of S/N, which is a considerably smaller factor than that by which N is observed to be depleted with respect to C. However, the reaction between H_2S and NH_3 would not remove *all* H_2S from the atmosphere, and there is spectroscopic evidence that H_2S is too scarce in the atmosphere today for it to have accounted for much of the ammonia depletion.

There is an argument, based on the observed abundance ratio in the Uranian atmosphere of deuterium (^2H or D) to the common hydrogen isotope (^1H), that the rate of exchange of icy materials between the mantle and the atmosphere is much lower than would be the case if Uranus were moderately hot and convective. This can account for the depletion of ammonia in the atmosphere because ammonia is slowly lost through photodissociation, and if its replacement from the mantle is slow then it will become depleted in the atmosphere. However, it is more likely that Uranus is moderately hot and that it is depleted *throughout* in ammonia. How could this have come about?

If the temperature of the region of the PFM from which Uranus formed was over 1000 K then most of the ammonia would be dissociated to yield hydrogen and nitrogen. A similar outcome would result if there were far higher intensities of solar uv radiation than exist today. Provided that the temperatures were not greatly in excess of 1000 K then little of the methane and water would be dissociated. However, a PFM hot enough to dissociate ammonia is too hot for the other ices to condense and therefore Uranus would not acquire much icy material. But if the PFM cooled then Uranus would acquire copious amounts of icy material. Moreover, if the cooling was fairly quick then little of the ammonia would reform, and if much of the hydrogen, helium and nitrogen in the PFM was then removed from the Uranus region then there would be little left for Uranus to subsequently capture, and thus Uranus would be depleted not only in hydrogen and helium but also in nitrogen.

Another possibility emerges by analogy with methane, as follows.

In the Uranian mantle, where the temperatures may be *several* 1000 K (Figure 12.3), methane is not significantly dissociated. This is because of the high pressure: you learned in section 2.1.4 that the higher the pressure the greater the tendency for chemical reactions to *minimize* the number of molecules. Thus, CH_4 is preferred to $(C + 2H_2)$. However, the *temperatures* near the centre of Uranus may be sufficiently high for the disruptive tendency of temperature on CH_4 to overcome the tendency of high pressure to reduce the number of molecules. In this case the carbon liberated would sink to the core and the H_2 would rise into the atmosphere. This tends to slightly decrease the C/H_2 ratio.

A similar argument would apply to any ammonia present except that both hydrogen *and* nitrogen would rise into the atmosphere. Therefore, it is possible that early in Uranus' history the ammonia became nitrogen and hydrogen in the atmosphere, the nitrogen, as in the cool PFM, only recombining *slowly* to form NH_3. Moreover, the dissociation of ammonia would happen at lower temperatures than for methane and therefore far less methane would be dissociated. Were there a far more massive atmosphere than today, and were it largely removed at about this time, then this too would account for Uranus' depletion of hydrogen, helium and nitrogen, though it is not clear how such a removal could occur.

12.3 Summary

Uranus is the first planet to have been discovered in recorded history, though it is so far away that it has not yet been visited by spacecraft. It is intermediate in mass between the giant planets and the terrestrial planets, and though its atmosphere consists largely of hydrogen it is depleted overall in hydrogen compared to the giants. By contrast, Uranus and the giants probably contain roughly comparable amounts of icy and rocky materials. Uranus probably has a moderately hot interior, and its cooling time may have been considerably lengthened by solar choking, which today may be blocking much of the heat that would otherwise be flowing from the interior. Heat from accretion and/or from subsequent differentiation are the likely sources of its internal heat. The N/C ratio in Uranus may be considerably less than the solar relative abundance of N to C, and various plausible explanations exist for this. The large axial inclination may be the result of the collision of a massive body with Uranus soon after it was formed.

12.4 Questions

1. How likely is it that Uranus could have remained undiscovered for several decades after 1781?
2. Outline the evidence that Uranus has a large axial inclination.
3. State which of the following ratios for Uranus are likely to be consistent with the solar relative abundances of the corresponding elements.
   (a) Uranus' atmosphere: He/H_2; C/H_2; N/H_2; C/N; He/C.
   (b) Uranus' interior: $(He + H_2)/(icy\ materials)$; $(He + H_2)/(rocky\ materials)$; C/H; C/N; icy materials/(rocky materials).
4. Why is the *atmosphere* of Uranus enriched in carbon compared to the solar relative abundance of C/H_2?
5. If there are methane clouds deep in the atmosphere of Uranus, but no clouds of ammonia, what effect does this have on the C/N ratio of the whole atmosphere compared with the observed ratio above any such clouds?
6. List the external observations of Uranus and any other evidence that has helped to constrain models of the interior. Indicate how the observational constraints could be made tighter.

7. How can a hot interior of Uranus be reconciled with the observation of little or no excess radiation?

8. Give a *one line* summary of each of the possible reasons for any ammonia depletion throughout Uranus.

13

THE SYSTEM OF NEPTUNE

13.1 The discovery of Neptune

In the midst of the French Revolution in the turbulent Paris of 1789 the French astronomer Jean Baptiste Joseph Delambre (1749–1822) was preoccupied with the orbit of Uranus.

Delambre had available numerous observations of Uranus spanning the period from the discovery in 1781 up to 1788, and also the two pre-discovery observations of Uranus that by then had been unearthed. The earlier of the two was in the records of the British astronomer John Flamsteed (1646–1719) and was made in 1690. The latter was made in 1756 by the German astronomer Johann Tobias Mayer (1723–1762). Of course, neither of them had suspected that that particular point of light they had recorded was a new planet.

The two pre-discovery observations were of particular importance in 1789 because they greatly extended the period over which positions of Uranus were known. Consequently, the orbital elements could then be calculated more accurately than with the post-discovery observations alone. The orbital elements were calculated from *some* of the observations, using Newton's Laws of Motion and Law of Gravity. Allowance was made for the perturbing effects of Jupiter and Saturn using their orbits and masses, and again using Newton's Laws. The result is a set of mean orbital elements from which it should be possible to calculate positions of Uranus that match *all* the existing observations to within the observational uncertainties. This Delambre was able to achieve. Furthermore, tables can then be prepared predicting the *future* positions of Uranus. Delambre published such tables in 1790.

This happy agreement between observation and theory lasted until 1820. It would have lasted *less* long had not Europe been in political turmoil for much of the intervening 30 years, and had not the astronomical community largely lost interest in Uranus. This loss of interest stemmed partly from the inevitable passing of novelty, and partly from the growth of interest in the apparently empty gap between Mars and Jupiter, where the Titius-Bode law, bolstered by the support given it by the discovery of Uranus at about the "right" distance from the Sun, predicted there should be a planet.

By 1820 there were observations of Uranus spanning nearly 40 years since its discovery, and in addition the number of pre-discovery observations had grown to seventeen. In that year the French astronomer Alexis Bouvard (1767–1843) attempted to correct his own tables of the positions of Jupiter, Saturn and Uranus, the position of the latter being particularly far from his predictions. He was also aware that the positions of Uranus predicted by Delambre differed markedly from the more recent observations.

Bouvard was able to establish orbital elements for Jupiter and Saturn that fitted all the observations. But he was *unable* to find a set of elements that fitted all the observations of Uranus. One set fitted the older observations, another set fitted the more recent observations. Bouvard resorted to rejecting the older observations as inaccurate, even though this meant attributing errors of about 50″ to astronomers who were generally regarded as having attained observational accuracies of about 5″.

Bouvard published his new tables of Uranus in 1821, and even though he had rejected the older observations he was still only able to match the more recent observations to within 9″, whereas the observational uncertainties at that time were only about 5″.

Bouvard's cavalier rejection of the older observations was not accepted by all astronomers. And alas! by 1825 the discrepancy between Bouvard's predicted positions and the observed positions of Uranus began to grow. By 1832 Uranus was nearly half a minute from its predicted position. Though this is only ⅟₆₀ of the angular diameter of the Moon it was *far* more than the discrepancy for any other planet.

Various solutions were suggested for the problem of Uranus. Among these were that Newton's Law of Gravity no longer applied beyond Saturn. Another claimed that there was one or more unknown bodies in the outer Solar System, and one of the first documented suggestions that there was a trans-Uranian planet came from the English amateur astronomer, the Rev. Dr. Thomas John Hussey, in a letter in 1834 to George Biddell Airy (1801–1892) then the Plumian professor of astronomy at Cambridge University. Hussey even suggested that he, Hussey, should undertake a search, but Airy discouraged him perhaps because of Airy's belief that the solution lay in modifying Newton's Law of Gravity.

One of the central figures in the discovery of Neptune is John Couch Adams (1819–1892), the son of a Cornish farmer. His talent for mathematics became apparent at an early age, and in 1839 he gained admission to Cambridge University to study mathematics. He graduated brilliantly in 1843 by which time he had become interested in the problem of Uranus and was convinced that the solution lay in the existence of a trans-Uranian planet. James Challis (1803–1882), Plumian professor of astronomy since Airy had become Astronomer Royal at Greenwich in 1835, gave Adams some encouragement and by October 1843 Adams had shown that it was *possible* for a trans-Uranian planet to account for the observed motion of Uranus. By

September 1845 Adams had calculated a mass and orbital elements for such a planet and he had showed that it could account for *all* the known observations of Uranus including the pre-discovery observations. The main assumptions that Adams had made were that the semi-major axis of the planet was 38.4 AU in accord with the Titius-Bode law and that the inclination of the orbit was negligibly small.

Adams showed his calculations to Challis and then left Cambridge to visit his family in Cornwall, leaving Challis with a position for the new planet for 30 September 1845. Adams had also pointed out that the planet should have a disc large enough to be seen with the biggest telescope at Cambridge, the Northumberland refracting telescope which had a main lens of 298 mm diameter. Challis made no search.

On his way to Cornwall Adams visited Greenwich to present Airy with a summary of his calculations, but Airy was in Paris. On his return from Cornwall Adams again called at Greenwich, but Airy was out and so Adams left a summary of his results. Adams returned to Airy's house later in the day but was turned away by the butler. It seems that Adams felt slighted.

Adams later heard from Airy that he was unenthused by the summary that Adams had left, and this probably explains in part why Adams to some extent lost interest in his trans-Uranian planet. There are several possible reasons for Airy's negative rection: Airy was probably still convinced that a modification of Newton's Law of Gravity was the correct explanation; he was *un*convinced that a theoretical solution along Adam's lines could be obtained with sufficient accuracy, indeed he was more inclined towards practical things and away from theoretical work altogether; and he distrusted youth.

In December 1845 there reached Airy a paper that had been presented to the French Academy of Sciences on 10 November. This paper contained an extremely thorough analysis of the motion of Uranus based on the observations made between 1781 and 1845. It, incidentally, discredited the work of Bouvard showing it to contain many errors. But it did not reconcile the motion of Uranus with theory. For example, it showed that orbital elements based on the observations made between 1790 and 1820 placed Uranus over 40″ from its actual position at the 1845 opposition. The paper concluded that the discrepancy

"can be attributed to outside causes whose effect I will evaluate in a second Memoir".

Airy regarded this paper as

"a new and most important investigation"

and that

"the theory of Uranus was now, for the first time, placed on a satisfactory foundation."

The author of the paper was Urbain Jean Joseph Le Verrier (1811–1877).

Le Verrier, son of a civil servant, was born in Normandy. He showed some early talent for science and finally gained admission to the Ećole Polytechnique in Paris in 1831. He seemed set for a career as a chemist when a good appointment in astronomy became available to him at the Ećole, which he took up. In those days there were few barriers between the different scientific disciplines.

At first he spent a good deal of time on studies of the orbits of comets, a subject which had also occupied Adams. But in the summer of 1845, after a small amount of earlier work, he was encouraged to concentrate on Uranus, and did so. The paper of 10 November 1845 was the first fruit.

On 1 June 1846 Le Verrier presented his second paper on Uranus to the French Academy of Sciences. In this paper he had calculated a mass and orbital elements of a trans-Uranian planet, and he gave its position in the sky for 1 January 1847, which he estimated was accurate to about 10°. The main assumptions, like those of Adams, were that the orbit had low inclination and that the semi-major axis was 38 AU, in accord with the Titius-Bode Law. However, he gave some additional reasons for adopting the distance, perfectly sound reasons which indicated that the planet should lie an appreciable distance beyond Uranus but not *very* far beyond Uranus.

When Airy received this second memoir in late June he could not "sufficiently express the feeling of delight and satisfaction which I received from it". He probably also noticed that Adams' and Le Verrier's positions for the new planet agreed to within 1°. Nevertheless, when he wrote to Le Verrier on 26 June he did not mention Adams.

It is difficult after so long to understand why Airy behaved in the way that he did. It could be that he now realized that with two solutions in such close agreement there may be something in the idea of a trans-Uranian planet after all, and that if Le Verrier was kept ignorant of Adams' work then perhaps Le Verrier would not press French astronomers to search for the planet, and the discovery might then fall to England. However, this does not explain why Airy was so *encouraging* to Le Verrier.

It was certainly not long before Airy pressed Challis and others to set up a systematic search for a trans-Uranian planet. Airy did not want to involve his own observatory, perhaps because the largest telescope was a refractor with a main lens diameter of only 170 mm, and perhaps because he did not want to disturb the large amount of routine work carried on there. Moreover, Airy was an old Cambridge man and he had been responsible for setting up the Northumberland telescope there.

Challis does not seem to have been particularly enthusiastic to embark on the search. This may have been because of personal and professional conflicts with Airy, coupled with Challis' scepticism regarding the existence of the planet in spite of his erstwhile encouragement of Adams.

Airy became concerned at the lack of action, and pressed Challis to start.

On 29 July 1846 Challis began his search. He started in a region of the sky where Adams had recently told him the planet should be found. The method Challis adopted was to record the position of every star and check each against a star chart: a "star" where none existed on the chart might be the planet. At that time this was a natural strategy because most observational work was on stars and on asteroids, objects which appear as *points* of light even at the highest magnification. However, Adams had urged Challis to use higher magnification and search the area for a *disc*: high magnification carries the penalty of covering only a small area of sky, but a disc can be readily picked out from a multitude of stars. Challis ignored Adams' suggestion.

By the end of August 1846 Adams had further refined his calculations and had obtained a new mass and orbital elements. By then Challis had covered the area of sky in which Adams had previously predicted the planet should lie. Challis had not checked all of his observations, but none of those he had checked had revealed a "star" where none should be. It seems that at this stage Adams' new position, which did not differ greatly from the earlier ones, did not inspire Challis to complete the checks.

On 31 August 1846 Le Verrier published his *third* memoir on his trans-Uranian planet in which he calculated a mass and orbital elements with even greater accuracy than before. He had also recalculated the position for 1 January 1847. The mass was 35.8 M_E, and the orbit is shown in Figure 13.1. This Figure also shows the August 1846 orbit of Adams, and the

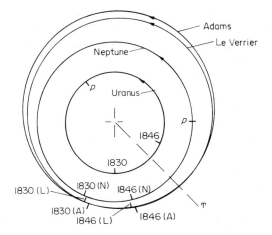

FIGURE 13.1

The orbit of Neptune and the predicted orbits of Adams and of Le Verrier. The viewpoint is perpendicular to Neptune's orbit, though the difference between the orbital inclinations of Uranus and Neptune is sufficiently small that the orbit of Uranus would not look very different were it viewed perpendicular to its own plane.

corresponding mass is 49.9 M_E. The masses are thus very similar, and you can see from Figure 13.1 that as viewed from the Earth in 1846 the directions in the sky of the two predicted planets are also very similar. Le Verrier had also predicted that the planet should have a disc large enough to be seen in Earth-based telescopes, and he estimated that at the predicted opposition of 19 August 1846 the planet had had an angular diameter of 3.3″. This estimate, and a similar one by Adams, was obtained from the predicted mass of the planet by making reasonable assumptions about its density.

Le Verrier had even less success persuading French astronomers to search for the planet than Airy and Adams had had in England, where apart from Challis there were certainly no more than one or two others making a serious search, and perhaps none others at all.

Le Verrier turned to Germany. On 23 September 1846 a letter from Le Verrier reached Johann Gottfried Galle (1812–1910) who was an assistant at the Berlin Observatory. Galle quickly persuaded the director (Johann Franz Encke (1791–1865) to allow him to search for the planet, and that same night he began, assisted by a young student astronomer Heinrich Louis d'Arrest (1822–1875). Using the Fraunhofer refractor, which had a main lens diameter of 229 mm, Galle searched the area of sky where from Le Verrier's orbit he had calculated the planet should lie that month. At Le Verrier's suggestion he looked for a disc, but found none. But then they checked all the stars they could see against a star chart, Galle at the telescope, d'Arrest at a new (1845) chart of that area of sky. Only a few had been checked when d'Arrest exclaimed that Galle had called out a star that was *not* on the chart. And it was only 55′ from the predicted position of the trans-Uranian planet. They watched it until it set at 2:30 a.m. Was it the planet? They picked it up on the following night: it had moved roughly the right amount (3″ per hour) against the stellar background. And careful scrutiny at a magnification of ×320 showed that it had a disc only a little less than the 3.3″ predicted by Le Verrier. The planet was about the size of Uranus.

Neptune had been found.

It was some months before the name was decided. Neptune in mythology is the son of Saturn and ruler of the remote ocean deeps.

By an ironic coincidence Airy was in Germany at the time of the discovery, and therefore heard of it several days before news reached Britain. It was not until 1 October, a week after the discovery, that the news finally reached Britain when *The Times* published a letter to the editor from John Russel Hind (1823–1895), director of the Bishop Observatory in Regent's Park, London.

It is hard to imagine the agony experienced by Challis on that October day. Just two days earlier, on 29 September, he had seen Le Verrier's third memoir and decided at once to search for a disc. That night he saw an object that seemed to have a disc but decided to check it another day. On 1 October, after he had read of the discovery, he did the check, and it *was* Neptune. But it was then too late to claim an independent discovery. More agonizing yet, Challis found that he had recorded Neptune on 30 July, but in checking that area of sky against a star chart he had stopped ten stars short of the planet. He had recorded it again on 4 August, but again that area of sky had not been compared against a star chart.

In early October Challis made public his months of searching for Neptune, but curiously did not mention Adams. It was John Herschel (son of William), one of the few other people who knew of Adams' work, who, on 3 October, finally announced it. This at once fuelled French suspicion that British astronomers were trying to muscle in on a French discovery. Why, it was asked, had not Challis, and Airy when he had written to Le Verrier in July, mentioned Adams? The French press became fairly hysterical, and the contemporary political tensions that existed between Britian and France were drawn into the issue.

Gradually, the storm subsided. Adams and Le Verrier met, became friends, and remained so until Le Verrier's death in 1877. And within a few years of 1846 it was widely acknowledged that Adams and Le Verrier were *co-predictors* of Neptune, though Galle and d'Arrest are regarded as *sole discoverers*.

Figure 13.1 also shows the *actual* orbit of Neptune. You can see that it differs considerably from those of Adams and Le Verrier. Moreover, the actual mass is only 17 M_E. This has led one or two astronomers to suggest that the discovery of Neptune was a happy accident.

This is *not* so. The orbital elements of Adams and Le Verrier place their predicted planets fairly close to Neptune during the first few decades of the nineteenth century, as Figure 13.1 shows. You can also see that Uranus was fairly close to Neptune during this period, and it was from Neptune's effect on Uranus that the existence of Neptune was deduced. It is therefore no accident that the orbital elements of Adams and Le Verrier place their planets in roughly the right *direction* during this period. The larger error in the predicted *distances* between Neptune and Uranus resulted in the predicted masses of Neptune being too large. Adams and Le Verrier had placed Neptune too far away because of their reliance on the Titius-Bode law.

When the *actual* orbital elements of Neptune were known it was checked whether Neptune could account for the discrepancies in the motion of Uranus. It could, to within the observational uncertainties of a few seconds of arc that then prevailed. These checks also yielded the first estimate of the true mass of Neptune.

Thus, in an era before the mechanical computer, much less the electronic computer, and with far fewer observations and with far larger observational uncertainties than today, Adams and Le Verrier had used Newton's Laws to *predict* the existence of a planet and had got the position roughly right thus leading to the planet's discovery. This was in the face of considerable scepticism that such a thing was possible, and such scepticism had reduced the urgency with which searches were set up and conducted.

The successful prediction of the existence of Neptune soon led to attempts to account for the residual precession of Mercury's perihelion by means of a planet or planets within the orbit of Mercury. However, in this case, as I outlined in section 7.0, it *is* a departure from Newton's Laws that provides the explanation. Airy would have been right about Mercury.

Many pre-discovery observations of Neptune are now known. The earliest is by Galileo. In December 1612 and January 1613, whilst he was observing Jupiter and its newly discovered satellites, he recorded a "star" which moved. This was Neptune. I wonder what the effect would have been on the history of astronomy and of science had he correctly identified the moving star?

13.2 The Neptunian system

Some consolation for British astronomy came on 10 October 1846 when William Lassell suspected that he had discovered a satellite of Neptune. By October 1846 Neptune was well past opposition, and therefore it was not until July 1847 that he was able to confirm its existence. It was named *Triton*. Its orbital elements enabled the mass of Neptune to be calculated far more accurately than from Neptune's effect on Uranus.

Lassell made his discovery in Liverpool with a recently completed reflector which had a main mirror 610 mm diameter and was then the biggest telescope in Britain. Until recently it was believed that Lassell had been sent Adams' predictions of Neptune's position late in 1845, nearly a year before the discovery. However, it now seems that he probably knew nothing of Adams' work until *after* the discovery.

A second Neptunian satellite was not discovered until 1949, by Gerard Kuiper in the USA. It was named *Nereid*. In mythology Nereid and Triton are attendants of Neptune. The existence of a third satellite is suspected from observations made in May 1981 when Neptune very nearly occulted a star. Near to the closest approach the starlight dimmed for several seconds. A ring would cause two dimmings, unless the very edge of the ring had occulted the star. The presumed satellite would be too faint to be detected directly and so confirmation of its existence must await future occultations. In August 1980 a less marked dimming in the light of another nearly occulted star may indicate a fourth satellite.

Table 13.1 lists the satellites.

Nereid has the most eccentric orbit of any known satellite in the Solar System. By contrast, the orbit of Triton is very nearly circular. The plane of Triton's orbit oscillates rather in the manner of a spinning plate just before it falls flat, as illustrated in Figure 13.2. The inclination of the orbital plane

TABLE 13.1
The satellites of Neptune

No.	Name	Year of discovery	a/(1000 km)	e	i/°	Sidereal orbital period/days	Radius/km	Density/ (kg/m³)
1	Triton	1846	355	0.00	159.9	5.877	1800–2600	5600–1900
2	Nereid	1949	5560	0.75	27.6	360	110–240	?
3	—	1981	50?	?	?	?	> 45	?
(4?)	—	(1980)	(37)	?	?	?	?	?

Neptune, equatorial radius at 1 bar = 24 600 km.
Notes: a = semi-major axis: e = eccentricity of orbit;
     i = inclination of orbit of Triton with respect to Neptune's equatorial plane and of Nereid with respect to Neptune's *orbital* plane;
     ? = not known.

with respect to the equatorial plane of the planet remains constant but the line drawn perpendicular to the orbital plane sweeps out a cone as shown. This can be called *precession of the orbital plane* and is caused largely by the J_2 coefficient of the gravitational field of a planet. The greater the angle between the orbital plane of the satellite and the equatorial plane of the planet, the greater the cone angle (Figure 13.2) and the more readily the precession is observed. The *rate* of precession yields J_2, which I shall take up in section 13.4. The line PQ in Figure 13.2 is expected to be parallel to the planet's spin axis, and half the cone angle gives the angle between the planet's equatorial plane and the satellite's orbital plane. In the case of Triton, as you can see from Table 13.1, the motion is retrograde with the plane of its orbit making an angle of 21.1° (180°−159.9°) with respect to Neptune's equatorial plane. (Had this angle been a lot less the precessional effect would have been too small to be measured.) Neptune's axial inclination (with respect to its own orbit) is then 28.8°.

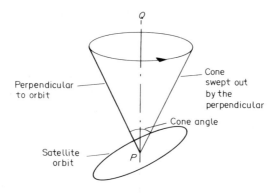

FIGURE 13.2
Precession of the orbital plane of a satellite.

Triton is so close to Neptune that it must be in synchronous rotation. It is within the orbit of Keplerian corotation and therefore tidal interactions must be causing the orbit to shrink. Moreover, it is so close to Neptune that its orbit must be shrinking so quickly that in 10 to 100 Ma Triton will pass within the Roche-limit and be tidally disrupted.

Triton thus seems to offer us the rare event of a cataclysm that, on a cosmic time scale, is to happen soon.

The mass of Triton has been determined from the motion of Neptune around the centre of mass of the Neptunian system. It can be shown that the separation of the two bodies and their orbital period yields the *sum* of their masses, and the position of the centre of mass yields their *ratio*. Thus, *both* masses can be found (compare section 3.2.1). The mass of Triton is 0.023

M_E, nearly double the mass of the Moon and about the same as the mass of Titan and Ganymede. The mass of Nereid is far less.

The radius of Triton has not been measured directly because its angular diameter is too small. Estimates have been made based on the amount of solar radiation scattered by Triton combined with plausible albedos, and on the amount of radiation emitted at planetary wavelengths combined with plausible emission rates per unit area. By these means a probable upper limit of 2600 km and a probable lower limit of 1800 km have been established for the radius. The respective densities are 1900 kg/m^3 and 5600 kg/m^3, and because icy materials were probably abundant in the vicinity of Neptune the lower density is more likely, indicating a radius nearer to 2600 km than 1800 km.

Spectrometric observations have revealed that Triton may possess an atmosphere containing methane at a surface pressure of about 10^{-4} bar. This is about the partial pressure expected of methane gas in equilibrium with solid methane at the dayside surface temperature of Triton, which is about 55 K. Any such surface ices have not been detected with certainty, but substantial contamination by rocky or carbonaceous materials would subdue the spectral signatures of solid ices. However, some astronomers dispute that even *gaseous* methane has been detected. Nevertheless, Triton could possess a substantial atmosphere of gases such as nitrogen with weak spectral signatures, though its greater distance from the Sun makes it unlikely that it has an atmosphere as massive as that of Titan.

There is tentative evidence that Triton is variable, possibly with a period equal to its orbital period, indicating, because of the likelihood of synchronous rotation, the existence of surface albedo features.

Very little is known about Nereid. A range of radii have been estimated from the amount of solar radiation it scatters combined with a range of plausible albedos. Its mass has not been measured.

No rings have yet been detected around Neptune, unless the third or fourth "satellites" are, in fact, rings.

No spacecraft have yet visited Neptune, but in September 1989 Voyager 2 is scheduled to fly by the planet, by which time Space Telescope should have been available in Earth orbit for a few years and should obtain pictures of Neptune not much inferior to those expected from Voyager 2. But for the time being Neptune remains a remote, poorly understood planet.

13.3 The atmosphere of Neptune

By much the same means as used for Uranus it has been established that Neptune possesses a hazy atmosphere consisting largely of hydrogen, with a roughly comparable enrichment of C/H_2 and a comparable depletion of N/H_2. Helium is expected to be present, but there are too few data as yet to infer its abundance. There seems to be more haze than in Uranus' atmosphere, and this could explain the weakness of the green tint imparted by gaseous methane.

Figure 13.3 shows an atmosphere profile for Neptune consistent with the (weak) constraints. The thermosphere is warmer than that of Uranus in spite of Neptune's greater distance from the Sun. This region of Neptune's atmosphere also contains more haze than the corresponding region of Uranus' atmosphere, and this probably explains the higher temperatures because haze particles are rather poor radiators at planetary wavelengths, and therefore solar radiation heats them up. The nature of the haze particles is unknown, though as on Uranus they are presumably derived from the atmospheric constituents with the aid of solar uv radiation. Nor is it known why Neptune has *more* haze than Uranus. A clear difference between the two atmospheres is that Neptune's atmosphere is more predominantly heated from below, the combined result of Neptune's greater distance from the Sun and greater amount of excess radiation, but whether this is the root of the explanation is not known.

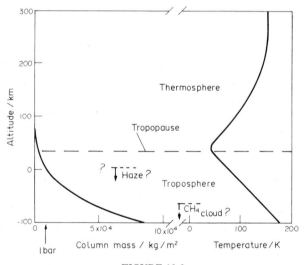

FIGURE 13.3
The vertical structure of the Neptunian atmosphere.

Neptune's greater distance from the Sun also means that ionization by solar uv radiation is less than on Uranus. Therefore, in any Neptunian ionosphere it is expected that ionization by cosmic rays will be about as important as ionization by solar uv radiation.

Beneath the haze there are probably one or more layers of cloud, though their altitudes and compositions are unknown. Methane and ammonia are possible candidates for cloud particles.

Features on the disc, as viewed at solar wavelengths, have been reported from time to time. A fairly persistent feature is an equatorial band that is slightly darker than the rest of the disc. There is evidence that this and other features arise largely from differences across the disc in the amount of absorption of solar radiation by methane. This could occur if the haze were thicker in some regions than in others, thus allowing different column masses of methane to be "seen".

The features on the disc vary with time. Short-term variations could be the result of changes in the haze content of the upper atmosphere though the reason for any such changes is unknown. There is also a longer-term variation in the amount of solar radiation scattered by Neptune. Unlike Uranus the changing aspect of Neptune does not seem to be the major cause. The major cause seems to be changes in solar activity, the *greater* the activity the *less* the amount of radiation scattered. This is also the case for Titan. In both cases it is likely that the changes in the intensity of solar uv radiation which accompany changes in solar activity cause changes in the amount or type of haze in the atmosphere, though the details are unknown.

13.4 The interior of Neptune

The mass of Neptune determined from the orbits of its satellites is 17.23 M_E, rather greater than that of Uranus. The best values for the equatorial and polar radii are obtained from a stellar occultation which occurred in 1968. The measurements, adjusted to the 1-bar pressure level, yield about 24 600 km for the equatorial radius, and about 24 000 km for the polar radius. The uncertainties in these radii are such that the flattening f could be anything from 0.017 to 0.031.

The mean density of Neptune is 1640 kg/m^3, significantly larger than the 1250 kg/m^3 of Uranus. This greater density in part arises from the greater mass of Neptune which gives greater compression of the interior. However, depending on the particular model of the interior adopted, there may be a small density difference between the *uncompressed* densities which then indicates slight differences in the relative abundances of the materials constituting the two planets.

Values of J_2, as outlined in section 13.2, have been obtained from the rate of precession of the orbital plane of Triton. This rate is difficult to determine accurately and therefore the observational uncertainty in J_2 is moderately large.

Attempts have been made to measure the axial rotation period by means of the same two techniques used on Uranus. Radiometric methods are slightly easier to apply to Neptune than to Uranus because the periodic variations in the intensity of the solar radiation scattered by Neptune are larger, presumably because of stronger albedo features. Such methods have recently yielded sidereal axial periods T_a around 19 hours. When the most reliable data are scrutinized then more than one period can be extracted. For example, over one particular time-span the periods 17.73 hours, 18.42 hours, and perhaps 18.56 hours seem to be present *simultaneously* throughout the time-span. This kind of result has been *tentatively* interpreted as evidence of an east–west wind system as on Jupiter and Saturn. On the two giant planets the longer periods are closest to the internal (System III) rotation period, and the same may be true on Neptune.

Spectroscopic methods have yielded values of T_a ranging from about 10 hours to about 18 hours. This large range arises from the great difficulty in making observations, even greater than for Uranus because of Neptune's greater distance from us. However, it is not known why these methods yield, on average, lower values of T_a than those from radiometric methods.

The spin axis orientation inferred from Triton's orbit is confirmed by those spectroscopic methods that in addition to T_a yield also the spin axis orientation.

The considerable uncertainties in T_a and in f lead to the same difficulties in constraining internal models of Neptune as in the case of Uranus (section 12.2.1).

Neptune differs from Uranus in that there is a clear radiation excess from Neptune, corresponding roughly to radiation to space at planetary wavelengths at about 2 to 3 times the rate at which Neptune absorbs solar radiation.

Non-thermal radio waves have *not* been observed, but this places no useful upper limit on the internal magnetic dipole moment. For much of its orbit Neptune probably lies outside the *Sun's* magnetosphere, the boundaries of which are determined by the *interstellar* wind. The bow-shock will be closer to the Sun than any tail, and so any Neptunian magnetosphere would be alternately exposed to the influence of the solar wind and the interstellar wind, presumably with consequences not seen elsewhere in the Solar System.

Figure 13.4 shows a model of Neptune constructed along very similar lines to those of Uranus and subject to a comparable degree of variation. However, the following broad features seem fairly certain.

The substantial radiation excess from Neptune makes it very likely that it has a moderately hot interior. It can be shown to be plausible that solar choking is far less efficient in the Neptunian atmosphere largely because of Neptune's greater distance from the Sun. This allows the internal heat to escape at a faster rate today than in the case of Uranus. A *small* contribution to the excess radiation may arise from tidal heating of Neptune's atmosphere by Triton.

It is likely that the mass fraction of icy materials lies between 50% and 80% and that the mass fraction of rocky materials lies between 20% and 35% the

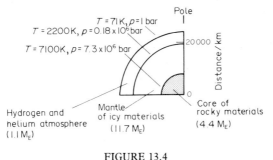

FIGURE 13.4

A model of Neptune.

greater the mass fraction of icy materials the smaller the mass fraction of rocky materials. It is also likely that most of the rocky materials are concentrated into a central core.

The compositional ranges are slightly different from those of Uranus (section 12.2.2). If you compare Figure 13.4 and 12.3 then you can see that for these two particular models Neptune contains a greater mass fraction of rocky and icy materials and a lower mass fraction of hydrogen and helium, and in these models it is this which largely account for Neptune's greater density.

The same remarks regarding the depletion of hydrogen, helium and ammonia apply to Neptune as apply to Uranus.

The rocky core in Neptune, because of the greater mass of Neptune and the correspondingly greater pressure on the core, may differ from that of Uranus in being *solid*. In this case it could possess no more than a small magnetic dipole moment. Water near the core may be sufficiently ionized by the high temperatures to yield a significant moment, but this is far from certain. In some models of Neptune there is sufficient hydrogen at sufficient depth to raise the possibility that such hydrogen is metallic but this is also uncertain. Thus, there exists the possibility that Neptune has only a weak internal magnetic dipole moment.

A plausible evolutionary sequence for Neptune, and one that is consistent with the model in Figure 13.4, is that Neptune became moderately hot during accretion 4600 Ma ago, or through subsequent differentiation, or by both means, and thereafter slowly cooled by radiating to space. Uranus and Neptune are sufficiently similar that by analogy much the same could well apply to Uranus. Thus, though Uranus is less readily demonstrated to have a moderately hot interior today, this does seem likely.

The evolutionary models indicate that Neptune (and presumably Uranus) never became as hot as Jupiter and Saturn. The underlying reason for this seems fairly clear and stems from the considerably greater masses of Jupiter and Saturn: you can imagine Jupiter and Saturn at the point where they are about as massive as Neptune and Uranus and then consider the gravitational energy to be released when the major part of their mass accumulates, largely in the form of hydrogen and helium. You can also imagine the greater amount of gravitational energy that would be liberated in the giants by any downward separation of icy and rocky materials, because of the greater distances involved in traversing these giants than in traversing the subgiants.

13.5 Summary

Neptune is the first planet whose existence was predicted by theory, and which was subsequently discovered.

Neptune is broadly similar to Uranus. The main differences are that its mass is 18.5% more than that of Uranus, that hydrogen and helium account for a rather smaller mass fraction, that its atmosphere is more hazy, and that its moderately hot interior is readily detected perhaps because solar choking is less efficient than on Uranus. The interior of Neptune is moderately hot today, and this is probably also true of Uranus.

Evolutionary models suggest that Neptune (and, by analogy, Uranus) was hot in the distant past, though not as hot as Jupiter or Saturn which would have been hotter because of their greater masses, and remain hotter today.

Triton is one of the most massive satellites in the Solar System. It probably has an atmosphere though it is probably not as massive as that of Titan. The demise of Triton may be at hand.

13.6 Questions

1. To what extent, if at all, was the discovery of Neptune a happy accident?
2. Why is Triton nearing tidal disruption?
3. Given that Triton and Titan started off with similar compositions, why is it likely that the atmosphere of Titan is more massive than that of Triton?
4. List the main differences between Uranus and Neptune, and in as many cases as possible state the possible reasons for the differences.

14

PLUTO, COMETS, ASTEROIDS
AND METEOROIDS

14.1 Pluto

In Chapter 13 you saw that after the discovery of Neptune the calculated positions of Uranus corresponded to its actual positions to within the limits of observational accuracy.

As time passed, observations of Uranus and Neptune accumulated, including pre-discovery observations of both planets. These observations enabled more precise determinations to be made of the orbital elements of Uranus and Neptune, and within a few decades it became clear that their motions could no longer be deduced from the gravitational influences of the known bodies in the Solar System.

Percival Lowell was one of a number of astronomers who used these discrepancies to predict the existence of a trans-Neptunian planet and on 18 February 1930 at the Lowell Observatory, and nearly 14 years after Lowell's death, a trans-Neptunian planet was discovered by the American astronomer Clyde William Tombaugh (b.1906). The planet was named Pluto.

Figure 14.1 compares the orbit of Pluto with one of two orbits, Orbit 1915B, which in 1915 Lowell had obtained for his trans-Neptunian planet, which he called Planet X. You can see that in 1930 the positions of Planet X and of Pluto were not very different, and it is tempting to think that Lowell, like Adams and Le Verrier before him, had predicted the existence of a planet from its gravitational effects on other planets, and that as a result this planet had been found.

Alas! this is not the case. First, Tombaugh discovered Pluto during the course of a *systematic* search of a band of sky straddling the ecliptic, and though Lowell's predictions led to this search being made the actual discovery was not guided by the predictions. Second, the mass of Pluto, or rather of Pluto plus its satellite Charon, is far too small to account for the discrepant motions of Neptune and Uranus. Refinements in data and in theoretical techniques, aided in recent decades by the electronic computer, have led to persistent downward revisions of the mass *required* of the Pluto system to the point where it now needs to have a mass of about $0.1\ M_E$. The

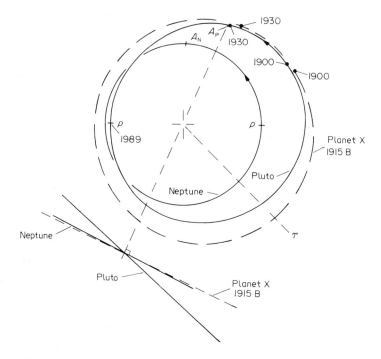

FIGURE 14.1
The orbit of Pluto and the B orbit of Lowell's Planet X. In the upper part of the
Figure the viewpoint is perpendicular to the orbital plane of Neptune. The
inclination of this orbit is given by Lowell as 10°, but the longitude of the
ascending-node is not given and therefore the B orbit has been placed in the
orbital plane of Neptune. The viewpoint in the lower part of the Figure is
edgewise to the orbits.

actual mass of the Pluto system was only established with reasonable accuracy
after the discovery of Charon in 1978, and the mass is a mere 0.0026 M_E,
which is about 40 times less than the required mass.

Therefore, Pluto is not Planet X after all. Moreover, searches have shown
that none of the trans-Neptunian planets predicted by various astronomers in
the early decades of this century exist. It seems that all of these predictions
are artefacts of limited data and of the substantial approximations that had to
be made in those days in performing the calculations.

Even the planetary status of *Pluto* is in question. Its mass is very small for a
planet, and its rather eccentric highly inclined orbit (Figure 14.1 and Table
1.3) also sets it apart. Its mass is more typical of the medium-sized satellites
of the outer planets, and so also is its predominantly icy composition,
indicated by a radius somewhere in the range 1300 km to 2000 km with a
corresponding mean density between 1500 kg/m^3 and 410 kg/m^3. Several

schemes have been suggested in which Pluto is an escaped satellite of Uranus or of Neptune, liberated by the passage of a body through the Uranian or Neptunian systems early in the history of the Solar System. Such an origin for Pluto could also account for its strange orbit.

It is also possible that Pluto could have accreted separately from any planet, and may be the most massive of a family of icy asteroids beyond Saturn. Another member of any such family may be Chiron (not Charon!) which is about one-third of the radius of Pluto and which occupies an eccentric orbit lying between Saturn and Uranus.

14.2 Are there any more planets?

The small mass of the Pluto system clearly leaves unexplained the discrepant motions of Uranus and Neptune.

There are a number of possible explanations. First, there may be a planet not far beyond Neptune with a very low albedo. Note that a very massive planet *way* beyond Neptune will not do because such a planet could not yield the *differences* in the discrepant motions of Uranus and Neptune, and moreover would also influence Saturn. Second, the comets (section 14.3) taken together may have sufficient mass. Third, there may be non-gravitational forces, such as would be caused by a tenuous gas. Fourth, there may be a small black hole not far beyond Neptune. Black holes are a theoretical possibility and in essence can be very small, very dense, and have zero albedo. Fifth, the icy asteroids about which I speculated in section 14.1 may be responsible. Sixth, the laws of gravity and motion may need changing, just as they were changed by Einstein. Seventh, there may be a planet in a highly eccentric orbit which was near enough to influence the motions of Uranus and Neptune in the decades before the various systematic twentieth-century searches were conducted, and which now lies very far beyond Neptune. Finally, there may be no discrepancies at all! Orbital motions can be established with great accuracy only when very accurate observations have been made over a substantial fraction of an orbital period. The orbital periods of Uranus and Neptune are 84 and 165 years respectively, and thus the very high accuracy of observations made in recent decades does not span the required fractions. However, should the recent and future observations show that there are no discrepancies then this could also be explained by an unknown planet in a highly eccentric orbit.

Some of these possible explanations give us good reason to search for trans-Neptunian planets. An independent reason is that a planet can lie as much as about 600 AU from the Sun and still remain in the Solar System, and at anywhere near such distances even a planet of Jovian mass would not influence Uranus or Neptune. This is also the case for planets not so far out but in highly inclined orbits. Searches could therefore be fruitful.

Tombaugh continued to search for planets until 1943, and established that within a wide strip of space above and below the ecliptic there are no planets brighter than the Earth would be at solar wavelengths were it removed to about 100 AU, and *probably* none brighter than the Earth would be at about 120 AU. A planet like Jupiter would probably have been detected out to about 470 AU. In addition to Tombaugh's extensive search several searches of much smaller areas of sky have been made, also without success. In 1976 the American astronomer Charles T. Kowal (b.1940) began an as yet fruitless search of a wide belt of sky around the ecliptic, and a few other extensive searches seem imminent. The existence of trans-Neptunian planets has been inferred from the orbits of comets (see section 14.3) but the arguments are not very convincing.

Recently a new method has become available for detecting distant planets, and this is via their gravitational influence on the paths of spacecraft which travel beyond Neptune. Pioneers 10 and 11 are the first to enter this domain.

But, for the moment, the outermost planet worthy of the name remains Neptune.

14.3 Comets

About 600 comets are known in the Solar System. They are small bodies, perhaps at most a few tens of kilometres in radius, and most of them move in large, eccentric orbits with a complete range of inclinations. Observations have established that water ice is the most abundant constituent of comets though there are appreciable quantities of other ices, and also silicates (possibly hydrated) and carbonaceous material.

When most comets are far from perihelion they are too faint to be seen. When such a comet moves inwards it will be seen first as a point of light moving against the stellar background. Most astronomers believe that at this stage the comet consists of a single solid body called the *nucleus*. The alternative view is that it consists of several solid bodies loosely bound together to form a *swarm*.

Within typically 3 AU of the Sun some of the comet's volatiles are evaporated by solar radiation to the extent that they form a dense atmosphere full of condensed droplets and dust particles which give the comet a fuzzy appearance. This fuzziness is called the *coma*, the Greek word for "hair", and this gives us the name "comet". As the comet approaches perihelion the coma can grow very large, in some cases to 10^6 km radius (0.007 AU). The coma and the remnant nucleus are called the *head*.

The growth of the coma is soon followed by the growth of two *tails*. These are tenuous streams of gas and dust lost by the coma, and which near perihelion can be seen extending up to 1 AU away from the head. Both tails can be seen in Figure 14.2, which shows Comet Mrkos, named after the

FIGURE 14.2
Comet Mrkos, imaged on 22 August 1957.

Czechoslovakian astronomer Antonín Mrkos (b.1918), and which he discovered as it moved through the night sky in the spring of 1957. The tails arise from the interaction of the coma with solar radiation and with the solar wind, both of which sweep coma material in a direction roughly away from the Sun. The tails differ in composition, the narrower filamentary tail consisting of ions, electrons, atoms and molecules, with a trace of *very* fine dust, the broader more featureless tail consisting largely of dust much of which is coarser though it is still only micro-sized.

As the comet recedes from the Sun the tail and then the coma shrink, and once again the comet becomes a faint point of light moving among the stars, in most cases vanishing as the comet recedes towards aphelion. The spectacular phase, when a comet is among the terrestrial planets, typically lasts only a few weeks.

A widely accepted theory of comets stems from work published in 1950 by the Dutch astronomer Jan Hendrik Oort (b.1900). He proposed that there are *hundreds of billions* of comets in a shell of space entirely surrounding the Sun and extending from about 10 000 AU to about 100 000 AU from the Sun, though the total mass of even such a large number of comets would not exceed, very roughly, 1 M_E. Gravitational disturbances by passing stars (the nearest star is presently 270 000 AU away) cause some of these comets to be ejected from the Solar System and others to be sent down among the planets where they move with a variety of orbital inclinations. Those that acquire perihelia less than about 3 AU become conspicuous enough for us to notice them. Most of these comets have aphelia well beyond Neptune, and therefore have orbital periods in excess of about 200 years. Indeed, about half are predicted to be in slightly *hyperbolic* orbits (section 1.1), which means that if they remain in such orbits they will escape from the Solar System after just one perihelion passage.

Of the 600 or so known comets about 500 indeed have periods in excess of 200 years, and about half of these are in slightly hyperbolic orbits. Moreover, the observed wide range of orbital inclinations is also predicted by the theory. Further support for Oort's theory comes from theories of the origin of the Solar System (Chapter 15) many of which predict the formation of numerous small icy bodies in the outer Solar System.

There remain the 100 or so known comets with periods less than about 200 years, and which are called *short-period comets* to distinguish them from the 500 or so *long-period comets*. The most famous short-period comet is Halley's Comet, named after the British scientist Edmond Halley (1656–1743). This has a period of about 76 years, the next perihelion being in 1986 (Table 1.2). The orbits of short-period comets are of fairly small inclination, and this holds the key to their relationship to the long-period comets, because a long-period comet in a low inclination orbit will have a relatively high probability of making a close approach to a planet. The outcome will *either* be the ejection of the comet from the Solar System *or* a reduction in the semi-major axis of the orbit, with a corresponding reduction in the orbital period. There will also be a reduction in the orbital eccentricity. Subsequent close approaches are likely to occur and the result is the conversion of a long-period comet into a short-period comet. In a sufficiently close approach a comet will be disrupted.

Each time a comet makes a perihelion passage within about 3 AU of the Sun it clearly loses some of its volatiles and also some of its smaller non-volatile particles. Thus, on average the short-period comets should be more depleted in these components than the long-period comets, in which case the comas and tails of the short-period comets should be the less prominent. That this is observed to be the case is further evidence for the orbital evolution.

Ultimately, the distinguishing coma and tail will be lost from short-period comets, in which case the remnants could in many cases become indistinguishable from certain types of asteroid (section 14.4) particularly since continuing orbital evolution places many of such comets among the asteroids.

The planets not only influence cometary orbits but must from time to time sweep up comets and cometary fragments. The contribution of the larger pieces to crater production on the terrestrial planets and on the satellites of the giant planets was estimated in sections 8.1 and 10.5. In the early history of the Solar System there could have been large numbers of comets moving among the planets, and this is one possible source of the late heavy bombardment, which in this case would have been post-accretional and would have affected all the planets.

14.4 Asteroids

The asteroids are small bodies, the great majority of which occupy prograde orbits which lie between the orbits of Mars and Jupiter. About 2000 have accurately known orbits and several thousand more have been glimpsed. Table 14.1 gives data on the three largest asteroids, Ceres (= "series"), Vesta and Pallas. The radii of the known asteroids extend down to a few hundred *metres*, the smaller asteroids accounting for most of the several thousand which have *not* had accurate orbits determined. Among the smaller asteroids the term "radius" should be taken to be *mean* radius, because in many cases they are irregularly shaped. Only the three largest asteroids have had their masses determined (Table 14.1).

TABLE 14.1
The three largest asteroids

No.	Name	Year of discovery	a/AU	e	i/°	Radius/ km	Mass/ $10^{-4}\,M_E$	Density/ (kg/m^3)
1	Ceres	1801	2.767	0.097	9.73	410–570	1.86–2.06	1200–3400
4	Vesta	1807	2.362	0.097	6.43	230–310	0.42–0.50	1800–4800
2	Pallas	1802	2.771	0.180	35.73	240–300	0.29–0.43	1700–3500

Notes: a = semi-major axis; e = eccentricity of orbit;
       i = inclination of orbit.

Figure 14.3 shows the number of asteroid orbits which have semi-major axes in the intervals shown. You can see that most orbits lie in a *main belt* in which the semi-major axes lie between about 2.2 AU and 3.3 AU. In Figure 1.1 it is this main belt range of semi-major axes which is shown at $n = 5$, plus the semi-major axis of the orbit of Ceres.

Outside of the main belt there are several small groups of asteroids, and most of these groups are shown in Figure 14.3. The Thules and the Hildas lie near orbital resonances with Jupiter, these resonances being indicated by ¾ and ⅔: for example, at ⅔ a body will orbit the Sun three times for every two orbits of Jupiter around the Sun. Other resonances seem to have depleted the main belt, and these are called *Kirkwood gaps* after the American astronomer Daniel Kirkwood (1814–1895). The Trojans (Figure 14.3) are divided between the two 60° Lagrangian points described in section 10.0, the two massive bodies being Jupiter and the Sun. The Atens, Amors and Apollos are collectively called *Earth-crossing asteroids* because they can have perihelia within the Earth's orbit and aphelia beyond it.

With few exceptions the orbital inclinations of the asteroids are less than 20° and the orbital eccentricities are less than 0.3.

Solar radiation scattered by asteroids brings to us various sorts of information on the basis of which asteroids have been sorted into different

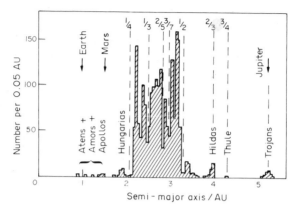

FIGURE 14.3

The distribution of semi-major axes among the orbits of the Asteroids.

groups. There are several different classification schemes, though it is possible that in certain cases nothing other than surface properties are being distinguished, because it is from the surfaces of the asteroids that the solar radiation is scattered. However, most astronomers believe that for most asteroids the nature of the surface indicates something about the nature of the interior.

The most important distinction is between those asteroids which are thought to consist largely of carbonaceous and rocky materials, and those which are thought to consist largely of metallic iron and silicates. I shall call these CR and IS types respectively. Among the *known* asteroids the number ratio of CR to IS increases with increasing distance from the Sun, and were the *unknown* asteroids to be included then this trend would surely increase, because the CR type are of lower albedo than the IS type and should thus be increasingly less liable to be discovered than the IS type as distance increases.

Asteroids must from time to time make sufficiently close encounters with each other to be placed in orbits from which they are either captured by the terrestrial planets or lost to the outer parts of the Solar System. Many of the Earth-crossing asteroids may be at a stage intermediate between populating the main belt and being captured by the Earth. Phobos and Deimos may be asteroids captured by Mars. Counteracting such losses there are thought to be gains, particularly from comet remnants.

The present total mass of known asteroids is about 0.0005 M_E. This is not very much but is nevertheless too much for the asteroid zone to be a "sump" which was initially empty and which has since captured material from elsewhere. It is also unlikely that the asteroids are the remnants of a disrupted planet, or rather of a *mini*-planet unless a good deal of asteroid material has since been lost. A major difficulty is that no plausible means have been suggested by which such a planet could have been disrupted. Further evidence against a disrupted planet origin is outlined in section 14.5.

Most astronomers believe that the asteroids are the result of arrested accretion. Some of these believe that this is because of the small and thinly dispersed amount of material in the asteroid zone, though this begs the question of why this zone has always been so deficient in mass. Others believe that Jupiter continuously prevents accretion in this zone by direct gravitational "stirring" of asteroids which increases their speeds relative to each other, whilst others believe that such "stirring" only happened a long time ago when Jupiter scattered a massive body through the asteroid zone. Such a body could also have removed a lot of material from the asteroid zone. Several other mechanisms have also been suggested.

This hypothesis of arrested accretion is in accord with the relative abundance trends across the main belt of the CR and IS type asteroids, because temperatures in the PFM would have been higher closer to the Sun and therefore *small* particles of material that condensed close to the Sun would have been depleted in carbonaceous material to yield the observed trend, preserved throughout the modest degree of subsequent accretion which yielded the asteroids. It can be shown to be unlikely that a uniform population of *asteroid*-sized bodies could have this trend produced across them: the trend has to be established among *small* particles.

14.5 Meteoroids, meteors and meteorites

Meteoroids are bodies in space smaller than a few tens of metres radius.

Roughly speaking the smaller the meteoroid the greater their number, and consequently, though the larger meteoroids cannot normally be seen in space from the Earth, the dust-sized ones are sufficiently numerous to give rise to a faint glow visible under favourable conditions. The larger ones can make their existence known when they pass into the Earth's atmosphere where they are heated to incandescence by friction, and blaze brief trails across the sky. Such meteoroids are then called *meteors*, and any sizeable fragments of a single meteor which survive passage to the Earth's surface are collectively called a *meteorite*. Numerous meteorites have been found, recognized by various non-terrestrial characteristics. Material which is vaporized and recondenses as a fine dust, or material originally present as fine dust, also finds its way to the Earth's surface where it can also be identified, particularly in polar ice and ocean sediments, largely through its non-terrestrial isotope ratios. In the course of a year you or I will have fall on us several milligrams of extra-terrestrial dust, though we do not notice it amongst the far greater quantity of dust of terrestrial origin. Overall, the Earth collects about 10^5 kg of meteoric material per day. This may seem a lot, but at this rate it would take 10^{15} years for the Earth to increase its mass by 1%.

There are various types of meteorite and correspondingly of meteoroids, ranging from *irons*, which, as their name suggests, consist largely of iron, to *carbonaceous chondrites* which consist of a mixture of hydrated silicates, carbonaceous material, and fragments of refractory and moderately refractory silicates. Carbonaceous chondrites bear no more than a trace of metallic iron. The main intermediate types are the *ordinary chondrites*, which consist largely of moderately refractory silicates, the *achondrites*, which in the main consist of silicates similar to those found in terrestrial igneous rocks such as basalt, and the *stony-irons*, which consist of roughly equal proportions of silicates and metallic iron.

The carbonaceous chondrites (CCs) are of great interest and importance. The relative abundances of the elements in the CCs, when allowances are made for the loss of the most volatile materials, are broadly similar to solar relative abundances, suggesting that they have never been subject to processes which produce differentiation. Moreover, the juxtaposition in CCs of materials of greatly differing degrees of volatility, and which in any case are not in chemical equilibrium with each other, indicates that the CCs as whole bodies have never been strongly heated. Radiometric dating of CCs shows that in many cases none of their components have been strongly heated since about 4600 Ma ago, which adds weight to this being the age of the Solar System. Moreover, there is an intricate argument based on oxygen isotope ratios in CCs which suggests that some of their material may have last been heated *before* the origin of the Solar System, and was thus not vaporized by that event. For all these reasons the CCs can be regarded as ancient, primitive bodies, and it is this which makes them interesting and important.

It is widely believed that the CCs have two sources, primitive asteroids and comets. A primitive asteroid is one which would have a broadly similar composition to the CCs, and like the CCs would not have been heated since it accreted. Some, or all, of the CR type of asteroid (section 14.4) could be such primitive asteroids. The CCs would then *either* be collisional fragments, or *very* small CR asteroids that need never have suffered fragmentation. The non-icy component of comets could also resemble CC material and thus provides a second source of CCs.

The cometary source is thought to give rise to *meteor showers*. A shower lasts from anything from a few hours to a few days, and in a heavy shower the number of meteors visible to the unaided eye is typically 100 per hour. Analyses of the paths of the meteors through the atmosphere show that the meteors in a particular shower moved through space in a tightly grouped cluster. Moreover, most showers recur annually, which suggests that the Earth moves through a stream of material spread around a common orbit around the Sun, as shown in Figure 14.4. Several of the annual showers occur when the Earth passes very close to the orbit of a known comet, and the shower is presumably the result of comet debris. Showers which do *not*

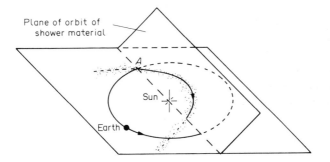

FIGURE 14.4
The Earth's orbit intersecting at A the debris-strewn orbit of a comet.

correspond to known comets probably arise either from completely fragmented comets, or from a devolatized comet too faint to be seen near
perihelion, or from a very long-period comet which has not passed through
perihelion in recent centuries.

Of the other types of meteoroid the ordinary chondrites are regarded as
the result of the heating of CC material, and the other main types are thought
to come from *non-primitive* asteroids, in particular from asteroids which have
become differentiated. Subsequent disruption of such asteroids could yield
irons from the interior, stony-irons from a transition zone, and achrondites
from the outer regions.

The modest amount of further evidence for these associations will not
concern us. However, it is important to note that those meteoroids which
seem to come from asteroids can be shown to provide strong support for the
view that the asteroids are *not* fragments of a disrupted planet.

Several attempts have been made to infer from the CCs the conditions that
existed very early in the PFM. However, there are many ways of producing
in space the various components of the CCs and of subsequently assembling
them into asteroids and, with icy materials, into comets. Therefore it is not
possible to use the CCs to *deduce* a unique set of conditions in the PFM,
though the presence of small particles of metallic iron in the CCs can be
shown to imply that CC materials were formed in a hydrogen-rich
environment. This indicates that the PFM itself was hydrogen rich, and this
is clearly in accord with the high abundance of hydrogen in the Sun and in
the giant planets today.

Attempts have also been made to infer from the other types of meteorite
the conditions that existed somewhat later in the PFM, but these attempts
also run into the difficulty that there are many different conditions that could
have led to the pertinent features of these meteorites.

Nevertheless, there is some hope that in the not too distant future the
study of meteorites will more firmly indicate the early history of the PFM.

14.6 Summary

Brief descriptions have been given of the smaller bodies in the Solar System which orbit the Sun separately from any planet. The largest of these is icy Pluto, which may be an escaped satellite of Neptune or Uranus, or the largest of a family of icy asteroids in the outer Solar System. The known asteroids are low-mass bodies, largely free of ices and which lie mainly between Mars and Jupiter. They are probably the result of halted accretion. The comets are low-mass icy bodies which probably originate in the outermost regions of the Solar System. We see only those members which travel through the inner Solar System, and many of these are ultimately devolatized and become virtually indistinguishable in their physical nature and in their orbits from certain types of asteroid. Meteroids are small samples of asteroidal and cometary material.

The relationship between comets, asteroids and meteoroids is summarized in Figure 14.5.

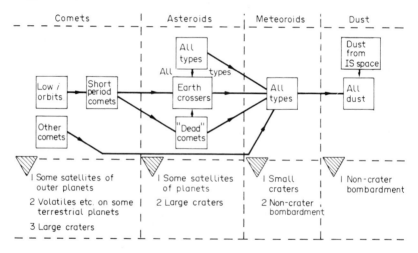

FIGURE 14.5

A likely evolutionary sequence of the smaller bodies in the Solar System. The downward pointing triangle ▽ indicates losses to the larger bodies in the Solar System.

14.7 Questions

1. Discuss the problem that was *thought* to have been solved by the discovery of Pluto, and outline possible new solutions.
2. Why is the planetary status of Pluto questionable, and what could it otherwise be?
3. What are the main structural features of a comet, and in what manner,

and why, do these features change (a) during a typical orbit, (b) during the comet's lifetime?

4. What are the main differences between the orbits of long-period and short-period comets, and how may the one type evolve from the other?
5. Outline the nature and the distribution of the asteroids.
6. What effects might Jupiter have had on the origin of the asteroids, and subsequently?
7. Explain those meteor showers which occur *twice* a year.
8. Summarize, in about 200 words, how the various types of meteoroid may have originated.
9. Outline the relationship between comets, asteroids, meteoroids and dust.

15

THE ORIGIN, AND END,
OF PLANETARY-SYSTEMS

15.1 The problem of the origin

It is not possible to *deduce* the origin of the Solar System from the state in which we observe it today, even if we knew everything about its present state. A large part of the reason for this is that matter and energy which was once localized within the Solar System has now spread out to great distances, and we have no means of identifying this matter and energy and of deducing when, in what form, and from where it left the Solar System. Therefore, theories have to be set up and then tested against various observational constraints, most of which are supplied by our (incomplete) knowledge of the Solar System.

The subject of the origin of the Solar System is called *cosmogeny*, and the theories of the origin are called *cosmogenic theories*. These terms can be extended to cover the origin of planetary-systems in general.

It is unreasonable to expect a cosmogenic theory to yield every detail of the present Solar System. There is only one way in which a theory could predict, for example, the diameter of the lunar crater Copernicus, or that you at this moment are reading these words, and this is by being as detailed atom by atom and moment by moment as the actual system. Our computational capacity and our knowledge is very far indeed from that required for this level of detail. Worse still, we cannot even demonstrate in rather broader terms the full potential of most theories. Nevertheless, a theory can be ruled out if it seems unable to create conditions which allow the possibility of your existence, or is unable to produce craters, or is unable to produce giant planets like Jupiter, or fails to produce any other broad feature of the Solar System.

There are, however, very many ways in which the broad features of our Solar System could have developed, and this is partly responsible for the large number of viable theories which exist at present.

Were other planetary-systems to be discovered then the number of theories could very probably be reduced. And were these other systems to be at various stages of evolution then, provided that there were good reasons to believe that the evolutionary processes in all of them are similar to those which operated in the Solar System, it might become possible to observe all the major stages through which our Solar System must have passed, in which case cosmogenic theories would have the comparatively straightforward task of linking these stages together. No such evolutionary stages have as yet been observed. However, it is likely that we have identified those environments in space which seem to be the most propitious for producing planetary-systems, and these are the *molecular clouds*.

15.2 Molecular clouds

Nearly all of the stars in the known Universe reside in vast assemblages of matter called *galaxies*. Figure 15.1 shows two galaxies which are roughly similar to the galaxy in which our Sun resides, and which I shall denote the Galaxy, with a capital "G". One of the two galaxies illustrates the edge-view of ours and the other the face-view. The face-view of our Galaxy is about 3.2 × 10^9 AU radius, compared to which the 30-AU semi-major axis of Neptune's orbit pales into utter insignificance. A common unit for measuring these vast distances is the light-year (LY), which is the distance that electromagnetic radiation travels through space in a year. The radius of the face-view of our Galaxy is about 50 000 LY. The Sun is about 34 000 LY from the centre, at a position equivalent to that marked in Figure 15.1. The Sun is one of about 10^{11} stars, of which there are a great variety, and it is such stars which collectively make a galaxy luminous.

For obvious reasons the type of galaxy shown in Figure 15.1 is named a *spiral galaxy*. About half of the very many known galaxies have this spiral form. These galaxies vary considerably in size and shape but the general nature of molecular clouds is probably much the same inside all of them. Molecular clouds might not be common in other types of galaxies because they contain far less gas and dust between the stars, which as you shall see is of central importance to our story.

The gas and dust between the stars constitutes the *interstellar medium*. It is not everywhere the same and there is little doubt that the most propitious regions for producing planets are those with the largest densities and the lowest temperatures. This is because such regions are the easiest to make collapse, and such collapse is necessary to produce dense objects like planets. Collapse is easier to achieve in such regions because the higher the density of a medium the greater the internal gravitational forces and the greater the tendency for it to contract to higher densities, and the lower the temperature the smaller the random speed of the molecules and the smaller the tendency of the medium to expand.

Stars are no less dense than planets, and therefore these same regions are also the most propitious for star formation. In this and in many other ways, as

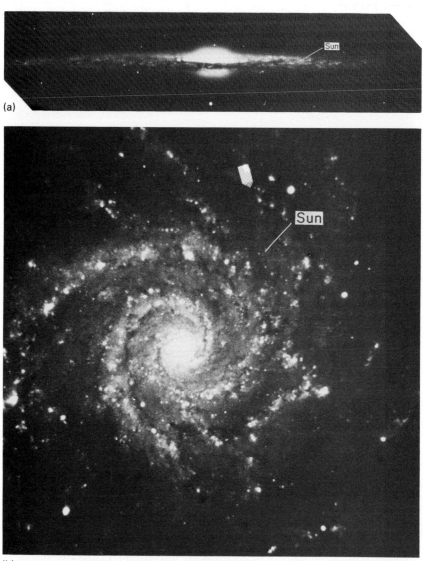

(a)

(b)

FIGURE 15.1
Similar galaxies to the Milky Way, showing where the Sun would lie (a) NGC 4565 (the edge-view), (b) M75 (the face-view). The Sun is about 34 000 light years from the centre of the Milky Way.

you shall shortly see, the origin of stars and the origin of planetary systems are connected.

In our Galaxy today the regions of space with the highest densities and the lowest temperatures are the molecular clouds. Moreover, though our Solar System is about 4600 Ma old the Galaxy is at least 10 000 Ma old, and for this and other reasons it is believed that the molecular clouds which we observe today are not substantially different from those which existed 4600 Ma ago.

The molecular clouds get their name from the predominance within them of hydrogen in *molecular* (H_2) form rather than the atomic (H) form which predominates the interstellar medium elsewhere. Molecular clouds vary in size from vast irregular structures with typical transverse dimensions of 300 LY and masses up to about 10^6 times the mass of the Sun ($M_\odot = 1.989 \times 10^{30}$ kg), down to more spherical clouds about 2 LY across and no more than a few tens of $M_\odot$ in mass. (Note that in this context $M_\odot$ is an appropriate mass unit because the Sun is a star of fairly average mass.) The temperatures in star-free molecular clouds are typically a few tens of Kelvins. The densities are conveniently expressed in terms of the number of hydrogen molecules per cubic metre, because H_2 accounts for nearly all of the mass. Molecular cloud densities range from about 10^8 to about 10^{12} H_2 molecules/m^3, and though these are high for the interstellar medium you should note that even 10^{12} molecules/m^3 corresponds to only about 0.003 kg of H_2 in a cube of side 10 km, which by laboratory standards is a very good vacuum! Nevertheless though molecular clouds account for no more than about 1% of the volume of our galaxy they still account for perhaps as much as 50% of the mass.

The high-density low-temperature conditions in molecular clouds are largely responsible for the predominance of molecules over single atoms, the most abundant example being H_2, and for a rather enhanced mass fraction of dust. However, though the mass fraction of dust is only modestly enhanced, to about 1%, the number of dust particles per cubic metre is considerably greater in molecular clouds than elsewhere in space because of the greater number density of all constituents. It is for this reason that molecular clouds can obscure the light from stars beyond the clouds, as illustrated in Figure 15.2. This Figure also shows the dust elsewhere in such clouds being lit up by stars.

The association of high density with low temperatures in molecular clouds is not merely a happy accident. Pressure increases as temperature rises and as density rises, and thus the conditions in a molecular cloud strike a rough pressure balance with the lower-density higher-temperature regions outside it.

15.3 Star formation

In most molecular clouds it is estimated that gravitational forces are sufficient to overcome the spreading effect of thermal motion, and therefore

FIGURE 15.2
The Horsehead Nebula (IC434), part of a molecular cloud which extends
beyond the lower edge of this picture. The area covered by this photograph at
the distance of this cloud is about 38 LY by 30 LY.

these clouds should be contracting. However, this is not the happy
conclusion that at first sight it would seem to be, because a number of
astronomers believe that the resulting rate of star formation in our Galaxy
would exceed that which observational evidence indicates. This gives us some
reason to seek ways of "stiffening" the molecular clouds to slow down the rate
of collapse. Magnetic fields, spin, and turbulent motions are all plausible
stiffeners. *Turbulent motions* are those in which a small region of a medium, in
this case a molecular cloud, moves more or less in one piece but in a different
direction from the motion of adjacent regions. However, as well as stiffening
a cloud turbulent motions can also *aid* contraction, because certain
"collisions" between regions would compress them to higher densities. Of
these three possible stiffeners spin is likely to be the most effective.

However, it may be that there is no appreciable stiffening of molecular
clouds, in which case a simple criterion for contraction can be applied. This
criterion was established by the British scientist James Hopwood Jeans

(1877–1946), and is based on the *Jeans mass*. This is the *minimum* mass that a medium of given composition, temperature and density must have before its internal gravitational forces can cause it to contract. If the Jeans mass of a medium is *less* than its actual mass then it will begin to contract. For a given composition, temperature and density it is clear that the larger the medium and thus the greater its mass the more readily the Jeans mass is exceeded. This has the very important consequence that it is *far* easier to form stars in a molecular cloud than to form planets, because stars are of far greater mass than planets.

The Jeans criterion also suggests that as a medium contracts fragmentation may occur. As contraction proceeds the density of the medium increases and it can be shown that its Jeans mass *decreases*. Ultimately, the Jeans mass will become considerably less than the mass of the medium and fragments of the medium will then have masses in excess of the Jeans mass. There will then be a tendency for such fragments to separate and thereafter to contract independently of each other. Fragmentation does not continue indefinitely, because the inward-moving matter picks up speed which is then partly lost in collisions thus increasing the random speeds of the constituents of the medium, and this manifests itself as a rise in temperature of the medium. This will slow down the rate of contraction. Moreover, it can be shown that a rise in temperature *increases* the Jeans mass and though this alone would halt fragmentation, spin and other effects may do so instead.

In a molecular cloud this fragmentation ensures that several less massive stars form rather than a single star of very high mass. However, fragmentation ceases well short of planetary masses and thus planets are not likely to be formed directly from a contracting molecular cloud. It is far more likely that planets are formed as adjuncts of star formation and thus nearly all cosmogenic theories include star formation as an integral part.

There is some observational evidence for the formation of stars in molecular clouds.

In our Galaxy there are numerous *star clusters*. Two examples are shown in Figure 15.3. Star clusters contain anything from about 100 to about 10^6 stars, and this corresponds to a mass range which covers all but the least massive molecular clouds. Moreover, the range of sizes of star clusters also corresponds to a size range which covers all but the smallest molecular clouds. This suggests that for each star cluster there was once a molecular cloud most of the mass of which has come to reside in the stars. In Figure 15.3 (b) the faint luminosity between the stars is suggestive of the last vestiges of free cloud material. Theoretical studies indicate that star clusters gradually disperse, and there are certainly star clusters which appear to be partially dispersed. Such dispersal could also account for the numerous stars which are not in clusters, such as the Sun.

Additional observational evidence for star formation in molecular clouds is

(a)

FIGURE 15.3
(a) The star cluster NGC5272. The area covered by this photograph at the distance
of this cluster is about 96 LY by 76 LY. (b) Part of the star cluster The Pleiades
(NGC 1432). The area covered by this photograph at the distance of this cluster is
about 9.7 LY by 7.7 LY.

provided by infrared studies of such clouds, which indicate the existence of
compact luminous bodies around which there is comparatively little cloud
material. These bodies could be *protostars*, that is, stars in the very early
stages of contraction. Moreover, some stars are at sufficiently shallow depths
in molecular clouds to be visible at optical wavelengths. Among the various
types are *T-Tauri stars* which seem to be surrounded by a tenuous medium of
gas and dust in violent motion, the *T-Tauri wind*. This can be interpreted as
material which has recently "boiled" off a recently formed star. It is not

(b)

FIGURE 15.3
(continued).

certain whether the Sun ever passed through a T-Tauri stage, because T-Tauri stars may be rather more massive than the Sun, but they do add to the evidence for star formation in molecular clouds.

The initial stage of star formation, the protostar stage, is poorly understood because much of the action is hidden deep in molecular clouds and probably happens rather quickly thus explaining why we have not yet managed to observe protostars at all sorts of stages of development. However, it is clear that a contracting cloud fragment would increase in luminosity from a low initial value as gravitational energy released during the contraction raises the temperature. Theoretical calculations indicate that this is followed by a fairly rapid gravitational collapse and therefore by a rapid reduction in surface area.

During the collapse the surface temperature of the protostar remains approximately constant, and therefore its luminosity falls.

This protostar stage extends up to time $t = 0$ on the right-hand side of Figure 15.4, which applies to a cloud fragment of mass 1 $M_\odot$. Note that $L_\odot$ denotes the present luminosity of the Sun. Such a graph of luminosity against surface temperature, is an example of a *Hertzsprung-Russell diagram*, named after the Danish astronomer Enjar Hertzsprung (1873–1967) and the American astronomer Henry Norris Russell (1877–1957).

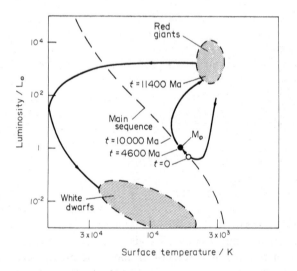

FIGURE 15.4
The evolution from birth to death of a star of about one solar mass.

At about $t = 0$, near the minimum luminosity of the protostar, the temperature near its centre will be high enough for the nuclei of hydrogen atoms, protons, to join together to make nuclei of greater mass, an example of a process called *nuclear fusion*. In the case of protons, nuclear fusion ultimately yields nuclei of helium. This releases energy and this halts the gravitational contraction of the star. The protostar consists largely of hydrogen, and this is therefore a prodigeous source of energy, and indeed is the dominant source of energy of a star until near its death. The onset of nuclear fusion marks the transition from the protostar stage to the stellar stage, and occurs about 10 Ma after the maximum luminosity attained during contraction of the cloud fragment. I shall return to the remainder of Figure 15.4 in section 15.6.

In most cosmogenic theories by the time a protostar has become a star there already exists an embrionic planetary-system.

15.4 Cosmogenic theories galore

Cosmogenic theories can be divided into those in which a star and its planetary-system form at about the same time and in a strongly coupled manner, and those in which this is *not* the case. These may be called *co-eval* and *non-co-eval theories* respectively. Co-eval theories may be subdivided into those in which the star and its planetary-system form the same body of contracting gas and dust, and those in which this is not the case, and these are called *nebular* and *non-nebular theories* respectively. Note that the *nebula* is the name of the gas and dust that comes to reside *outside* of the star: it does not include the star itself.

For the last few decades nebular theories have been far more widely supported than have any other category of cosmogenic theory. There are now numerous kinds of nebular theory but they all share the feature that a fragment of a molecular cloud somehow starts to contract, detaches itself from the rest of the cloud, and because whatever angular momentum it initially had is thereafter fixed its rate of spin increases as it collapses and it flattens out along its equatorial plane. The *differences* between the different kinds of nebular theory can be illustrated with thumbnail sketches of a couple of examples.

Low mass nebular theories start with a molecular cloud fragment which is only a few percent more massive than the mass of the Sun. As it contracts the few percent excess material is left behind in the equatorial plane of the fragment to form a thin, flat circular disk. The initial angular momentum is chosen so that the disk extends out to about the present orbit of Neptune. The disk is a representative sample of the molecular cloud and thus consists largely of gas (mainly hydrogen and helium) plus a small amount of dust. At this stage the Sun is a contracting blob at the centre of the disk, around which the disk material moves in roughly Keplerian orbits.

The dust settles gravitationally towards the disk's mid plane to form a thin sheet. Its density reaches a point where it becomes gravitationally "cohesive" and it breaks up into a large number of patches each of which gravitationally contract to yield numerous bodies up to a few kilometres in diameter. These are called *planetesimals*. Further accretion occurs via the capture of one planetesimal by another, and the outcome is eight major bodies. The inner four are Mercury, Venus, The Earth and Mars, the outer four are the *cores* of Jupiter, Saturn, Uranus and Neptune. The dust in the inner disk was devoid of icy materials, because of the proximity of the Sun. This dust forms the terrestrial planets which are thus made of rocky materials, including iron. In the cooler conditions further from the Sun icy materials also condense. Therefore, the cores of the outer planets have a rock plus ice composition, but more importantly, they become several times the mass of the Earth. Consequently they are able to gravitationally capture some of the copious quantities of hydrogen and helium in the disk.

Meanwhile, the Sun has developed a powerful solar wind and a copious uv flux, both of which begin to drive the disk gases away. By the time the cores of Jupiter and Saturn form, the disk is already depleted in gases, and this results in the observed modest depletion of hydrogen and helium in these planets with respect to rocky and icy materials. The cores of Uranus and Neptune form later, and this is why these planets are considerably depleted in these gases. By the time these subgiant planets have formed, a few tens of Ma have elapsed since the cloud fragment began to collapse.

The present slow spin of the Sun is the result of the interaction of the solar wind with the solar magnetic field. This transfers angular momentum to the wind, and most of this is lost from the Solar System.

High mass nebular theories start with molecular cloud fragments up to about two solar masses and with a large amount of angular momentum. The massive disk which consequently forms is sufficiently dense for some of it to gravitationally break up to form a small number of giant gaseous *protoplanets*. These contain all the components of the molecular cloud in roughly unaltered proportions. Cores form within these protoplanets and they undergo a variety of evolutionary sequences to yield the planets. Most of the gas in the disk is lost via the action of the turbulence which is to be expected in a dense disk, and which causes intense heating of the disk's surface, thus driving off gases. This turbulence is also largely responsible for transferring angular momentum away from the Sun.

Among the numerous other cosmogenic theories there is a non-co-eval theory which is largely the work of the British physicist Michael M. Woolfson (b.1927). He proposes that in a molecular cloud protostars form when two streams of material moving in opposite directions meet roughly head-on. Such streams arise from turbulence in the cloud. Their meeting compresses material to the point where the Jeans mass is exceeded and gravitational contraction ensues. A roughly head-on collision can occur, and this will compress enough mass by a sufficient amount to exceed the Jeans mass and such a head-on collision also yields a protostar with roughly the amount of angular momentum possessed by stars like the Sun. However, this small amount of angular momentum means that no disc can emerge from the protostar, which contracts to form only a star.

The material to make planets is drawn by the star from a passing protostar by means of tidal forces, the protostar being the one to lose mass because it is less compact and thus not as gravitationally bound as the star. Much of this material is captured into roughly coplanar orbits around the star, and this material subsequently evolves into a planetary-system. This is a *tidal-capture theory*.

These three theories have been articulated in considerably greater detail than that given in these thumbnail sketches, and all three yield many of the main features of the Solar System and also some of the more minor features. However, all three theories have to make some arbitrary assumptions, though future developments could obviate some of these.

The considerable differences between these three theories highlights the great difficulties and uncertainties which face cosmogeny. Astronomers are ignorant of many of the basic conditions in molecular clouds and of some of the processes of central importance to star and planet formation. In particular; the separation and collapse of a cloud fragment; the nature and roles of magnetic fields and plasma; the degree and role of turbulence; the initial angular momentum of a typical fragment; processes of accretion and of protoplanet evolution. *All* theories have to bypass this ignorance by making assumptions about many of these things, their plausibility often being questioned by others. Moreover, each theory has to make assumptions peculiar to itself, and each finds it difficult to predict certain features of the Solar System.

Cosmogeny will make progress through the further articulation of theories and by a deeper theoretical understanding of the important cosmogenic processes. However, progress could be spectacular if certain critical observations could be made. For example, if disc-like structures were to be observed in molecular clouds then this would be a great boost to nebular theories. A variety of critical observations could be made by new telescopes in space and even at the Earth's surface. Also of importance are developments in infrared techniques, because much of the radiation from molecular clouds lies at infrared wavelengths. A very interesting and useful discovery would be of ready-formed planetary-systems, and this is taken up in the next section.

15.5 Other planetary-systems, other civilizations

The great majority of cosmogenic theories predict that a high percentage of the stars in our Galaxy should possess planetary systems. Woolfson's theory is one of the few which predicts a small percentage, between about 1 in 10^3 and 1 in 10^7. Even in this case our Galaxy would have between about 10^4 and 10^8 planetary systems, and this is including only the 10^{11} stars which are luminous at present.

One can thus look to the stars with some hope of detecting other planetary-systems, and there are four main techniques by means of which this could be achieved. First there is *direct imaging*, though planets would be so faint and so near the far brighter star that the only hope for this technique is to observe nearby stars with large telescopes in space. Second, in the *occultation technique*, planets could be detected via the slight diminution they would cause in their star's brightness were they to pass between us and the star during their orbits. However, for any one star we would have to be observing a planet's orbit very nearly edgewise and during the very small fraction of its period during which the planet passes in front of its star. The third technique relies on the slight motion of a star around the centre of mass of the star and its planetary-system. By accurately measuring the position of a star it would be possible to detect this motion. However, this *astrometric technique* not only requires *very* accurate positional determinations but also requires long series of observations, because the star would have to make several orbits around the centre of mass before it could be established that any such orbital motion existed. Finally, the motion of the star around the centre of mass imparts periodic Doppler-shifts to the wavelengths of spectral lines in the radiation from the star, and these shifts could be detected, provided that the star's orbital motion were *not* viewed face-on. The motion would take as long to establish as in the astrometric technique but this *Doppler-shift technique* has the advantage that the size of the shift does not diminish with distance, and it can therefore be used at greater ranges and thus encompass far greater numbers of stars than the other three techniques.

So far none of these techniques has yielded unambiguous evidence of other planetary-systems. However, developments in the near future are likely to considerably enhance these techniques, and if this failure to detect planetary-systems were to persist then this would indicate that such systems are rare, and this would be of significance. For example, it would tend to support Woolfson's cosmogenic theory (section 15.4).

The most spectacular indication that other planetary-systems exist would be contact with a civilization from beyond the Solar System, because it is very unlikely that life could arise independent of a planetary-system. So far there is no good evidence for any such contacts. Nevertheless it is important to estimate the probable number of technological civilizations that exist in our Galaxy, or rather those which unwittingly or otherwise signal their presence across interstellar space. These I call *signalling civilizations*. It has been estimated that, on average, about one planet per planetary-system will be suitable for the development of life, and that life develops on about 1 in 10 of these. Woolfson's cosmogenic theory gives the *least* number of luminous stars in our Galaxy likely to have planetary-systems, about 10^4. Therefore there are probably *at least* 10^3 planets in our Galaxy on which life has developed. Estimates of the probability that any such life develops into a signalling civilization vary enormously, but 1 in 10^3 is among the *lower* estimates. Therefore, among the still luminous stars in our Galaxy there is a good chance that there has developed at least *one* signalling civilization other than ours. If planetary-systems are far more common than Woolfson's minimum estimate then this number would rise considerably.

Now we come to the *really* speculative bit, because even if a large number of signalling civilizations have *developed* in our Galaxy there could still be very few in existence today if it is a characteristic of such civilizations that they are short-lived. Our own civilization has only been unambiguously signalling its presence across interstellar space for about 60 years. If it were to end soon and if the Earth were never again to bear such a civilization then the Earth would have borne a signalling civilization for only about $\frac{1}{10}^8$ of the Earth's lifetime. If this is typical then there are probably *no* other signalling civilizations in our Galaxy today.

It is clearly very difficult to etimate the probability of future contact with an extraterrestrial civilization. There can, however, be no doubt that the answer to the question "are we alone?" is of enormous significance regardless of whether the answer is "yes" or "no".

15.6 The death of stars and of planetary systems

In contrast to the uncertainties of our origin there seems little doubt what awaits the Solar System, because observations of numerous stars at various stages of their life-cycles, plus well-founded theories of stellar evolution, have yielded a fairly clear and reliable picture of stellar evolution after the protostar stage.

In section 15.3 we left the Sun, or any other star of one solar mass ($M_\odot$), at $t = 0$ in Figure 15.4. This point lies on the dashed line labelled *"main sequence"*. This line runs through the $t = 0$ points of stars of different masses, from low masses at lower right to high masses at upper left. A star spends most of its life on the main sequence, and during this time its position in Figure 15.4 does not change very much from that at $t = 0$ though the luminosity of the star does rise by a few tens of percent, as shown for a star of mass 1 $M_\odot$, and this rise can be significant for the surfaces and atmospheres of planets, as you have seen earlier.

After a star of mass $1 M_\odot$ has spent about 10 000 Ma on the main sequence the drama of the star's death begins. The star begins to increase in size, and though its surface temperature falls a little, giving it a slightly red tint, its luminosity increases relatively rapidly. It leaves the main sequence and after a few hundred Ma it has become a very large red-tinted star of very much greater luminosity. It is then called a *red giant* (Figure 15.4). After a further 100 Ma or so it begins to contract and its surface temperature begins to rise, making it whiter to the eye, and after a further few tens of Ma it has become a *white dwarf*. During its transition to the white dwarf stage the star may fling off a small fraction of its mass. Then it cools and becomes a dense dark object. There are doubtless many such objects unseen in our Galaxy, some with the remnants of their planetary-systems around them.

In the case of the Solar System, as the Sun leaves the main sequence about 5000 Ma from now the temperature of the Earth's surface will start to rise and the oceans will begin to evaporate. Within a few million years the Earth may come to resemble Venus. As the luminosity of the Sun continues to rise the Earth will lose to space its atmosphere, oceans and other surface volatiles, and become a dry and atmosphereless world bearing scars long ago carved by weathering and by geological processes, particularly Plate Tectonics. There will be no life on the Earth and probably no evidence that life ever existed, except perhaps for artificial satellites still in Earth orbit.

By the time the Sun becomes a red giant it is certain that Mercury and Venus will have been engulfed: their orbits will lie inside the Sun and they will have evaporated. It is possible that the Earth will have suffered the same fate. By this stage Mars will have become devoid of its atmosphere and surface volatiles, and the remaining planets will have been greatly modified, much hydrogen and helium having been lost. Some of the larger satellites in the outer Solar System may become temperate. But any such havens for life will be short-lived. By the time the Sun becomes a white dwarf the surviving planets will be extremely cold and when the Sun becomes dark and cold the planets will be comparably frigid with surface temperatures of a few tens of Kelvin and with no significant interior sources of energy still alive to keep the interiors significantly warmer.

This general picture of the evolution of a star of one solar mass applies also to less massive stars and also to stars up to about $1.4M_\odot$, the more massive the star the more quickly it moves through its life-cycle.

For stars of mass greater than about $1.4M_\odot$, the shedding of mass between the red giant and white dwarf stages removes a substantially greater fraction of the star's mass than in the case of less massive stars. For stars with masses greater than about $4M_\odot$ the red giant stage is followed by a supernova explosion in which almost all the star's mass is dispersed into interstellar space. Such an explosion would vaporize all of any planets encircling such a star.

Nuclear fusion in stars creates many elements from the hydrogen "building block". Thus any mass loss from stars injects into interstellar space such elements, and supernovae are particularly important in this respect.

15.7 Summary

It is not possible to work back from the present Solar System to *deduce* how it was created. However, it is believed that the stars and any associated planetary systems are almost exclusively created in molecular clouds though there are insufficient observational data and many important processes are too poorly understood to be sure how this happens. As a result there is a great variety of cosmogenic theories though at present solar nebular theories are fashionable. There is some hope that improved observational techniques plus further theoretical work will considerably reduce the range of possible cosmogenic theories in the fairly near future.

In contrast to the obscurity of our origins our distant fate seems far more certain. Stars end in a blaze of activity, the more massive the star the shorter its life and the more spectacular the blaze. A good deal of the material incorporated into massive stars is returned to interstellar space with some enhancement of the elements other than hydrogen, including the elements which constitute interstellar dust. Thus the life-cycle of a star gives a new meaning to the funereal phrase "dust to dust".

The Sun will become a red giant in about 6000 Ma, by which time it will have incinerated the Earth. However, this cataclysm lies about three million times further in the future than the time of Christ lies in the past. Perhaps by then we will know how the Solar System was formed and whether other civilizations have ever existed in our Galaxy.

15.8 Questions

1. Outline the classification scheme of cosmogenic theories used here and add some extra possible categories of your own, with a few notes on each.
2. Why is the number of viable cosmogenic theories so large and how could the number be reduced?
3. For each of the three cosmogenic theories outlined in this Chapter list their common features, their major differences and the main difficulties that they face.
4. From the four techniques outlines in section 15.5, by means of which other planetary-systems may be detected, select the one most likely to be successful in each of the following cases:
   (i) a distant massive star with a planet in an orbit presented edgewise to us,
   (ii) a middle-distance massive star with a planet in an edgewise orbit,
   (iii) a nearby star,
   (iv) a distant star of low mass.
5. Outline the relationship between stellar evolution and the birth, life and death of a planetary-system.

FIGURE ACKNOWLEDGEMENTS

Figure no.	Source	Source ref. no.
2.1	1(c)	AS11-36-5352
2.12	7	—
3.2	1(d)	Viking-16735B
3.3 (a)	1(d)	Viking-18612
(b)	1(d)	Viking-17873
3.10	2	I-961
3.11	1(d)	Viking-17872
3.12	1(a)	Viking-17444
3.13	7	—
3.14	1(b)	Viking-17022
3.15	1(a)	Viking-45B60
3.16	1(d)	Viking-17690
3.19 (a)	1(a)	Viking-17138
(b)	1(d)	Viking-16983
(c)	1(d)	Viking-84A47
4.2	5	282
4.5	7	—
4.6	4	—
4.8	1(a)	Mariner X-14400
6.1 (a)	6	L3
(b)	1(a)	LO IV-187M
6.4	1(e)	AS17-145-22165
6.7	1(f)	AS17-151-23260
6.8	1(a)	Apollo 15-1557
7.4 (a)	1(a)	Mariner X-AOM18
(b)	1(a)	Mariner X-AOM19
7.5	1(g)	Mariner X-15427
7.6	1(b)	Mariner X-14855
7.7 (a)	1(a)	Mariner X-AOM11-27

Figure no.	Source	Source ref. no.
8.3	3	—
9.2	8	—
9.3	1(a)	Voyager-20993
9.4	1(b)	—
9.7	9	—
9.11 (a)	1(d)	Voyager-21282
(b)	1(d)	Voyager-21207C
(c)	1(d)	Voyager-21208C
(d)	1(d)	Voyager-21305C
10.3	1(d)	Voyager-22993
10.4	1(b)	—
11.2	1(a)	Voyager-23110
(inset)	1(a)	Voyager-23053
12.1	10	—
14.4	5	296
14.6	11	—
15.1 (a)	5	25
(b)	5	4
15.2	6	N44
15.3 (a)	6	D1
(b)	6	D4

Sources

1 NASA (a) From the US National Space Science Data Center via the US World Data Center A for Rockets and Satellites.
        (b) From JPL (USA).
        (c) From the US EROS Data Center.
        (d) From a commercial source.
        (e) From the US Lyndon B. Johnson Space Center.
        (f) From W. Kaufmann.
        (g) From J. Guest.
2 US Geological Survey.
3 Smithsonian Institution ©, Washington D.C., USA.
4 USSR Academy of Sciences.
5 Palomar Observatory photograph (USA) — via the Bookstore of the California Institute of Technology.
6 Lick Observatory photograph (USA).
7 Michael Kobrick.

8 Adapted from W. J. Kaufmann and E. J. Smith.
9 Adapted from Andrew Ingersoll.
10 Adapted from W. J. Kaufmann.
11 Adapted from B.Zellner and J. C. Gradie.

SOME FURTHER READING

I have included books which I have had a chance to examine and which I think could usefully be read either in parallel with this book or immediately afterwards. This excludes numerous research-level books and the even more numerous articles in research-level journals. However, I have included a few books which lie at a rather higher level than this book, and these are marked with an asterisk (*). I have also excluded numerous good books on astronomy as a whole, and on the Earth in particular.

General astronomy

All of the following are recent general texts with a good deal on the Solar System.

1. *Astronomy and cosmology*, by F. Hoyle, W.H. Freeman & Company, 1975.
2. *Essentials of astronomy*, by L. Motz and A. Duveen, 2nd edition, Columbia University Press, 1977.
3. *Exploration of the Universe*, by G.O. Abell, 4th edition, Saunders College Publishing, 1982.
4. *Discovering astronomy*, by W.H. Jeffreys and R.R. Robbins, John Wiley & Sons, 1981.

The Solar System — general

*1. *Interiors of the planets*, by A.H. Cook, Cambridge University Press, 1980 (mainly on the physics of planetary interiors).
2. *The new Solar System*, ed. by J.K. Beatty *et al.*, Sky Publishing Corporation, 1981.
3. *Pictorial guide to the planets*, by J.H. Jackson and J.H. Baumert, Harper & Row, 1981 (not much content except for the very good pictures).
*4. *Planetary exploration*, papers given at a 1980 Royal Society discussion meeting, The Royal Society, London, 1982.

5. *The planets*, by P. Frances, Penguin Book, 1981 (mainly on surfaces).
6. *Planets and moons*, by W.J. Kaufmann III, W.H. Freeman & Company, 1979 (mainly on atmospheres and surfaces).
7. *The Solar System*, by J.A. Wood, Prentice-Hall, 1979.

The Solar System — origin

1. *The cosmogeny of the Solar System*, by F. Hoyle, University College Cardiff Press, 1978 (Hoyle's recent views).
★2. *The origin of the planets*, by I.P. Williams, Adam Hilger, 1975 (a survey of many different theories).

The Solar System — terrestrial planets

1. *Earthlike planets*, by B. Murray *et al.*, W.H. Freeman & Company, 1981 (mainly on surfaces).
2. *The inner planets*, by C.R. Chapman, Charles Scribner's Sons, 1977.
3. *Planetary geology*, by J. Guest *et al.*, David & Charles, 1979 (mainly on surfaces).
4. *The realm of the terrestrial planets*, by Z. Kopal, The Institute of Physics, 1979.

The Solar System — some historical topics

1. *The discovery of Neptune*, by M. Grosser, Dover, 1979.
2. *Out of the darkness: the planet Pluto*, by C.W. Tombaugh and P. Moore, Stackpole Books, 1980.
3. *Planetary encounters*, by R.M. Powers, Sidgwick & Jackson, 1982 (mainly on the history of spacecraft exploration).
4. *A revolution in the Earth sciences*, by A. Hallam, Clarendon Press, 1973 (the history of Plate Tectonics).

The Solar System — special topics

★1. *Atmosphere and ocean*, J.G. Harvey, The Artemis Press, 1976 (the present state of the Earth's atmosphere and oceans).
2. *Atmospheres*, by R.M. Goody and J.C.G.Walker, Prentice-Hall, 1972 (the atmospheres of all the planets).
3. *Comets, readings from Scientific American*, W.H. Freeman & Company, 1981.
4. *Earth*, by F. Press and R. Siever, 2nd edition, W.H. Freeman & Company, 1978.

5. *The Earth: its birth and growth*, by M. Ozima, Cambridge University Press, 1981.
6. *The geology of Mars*, by T.A. Mutch *el al.*, Princeton University Press, 1976.
7. *Guide to Mars*, by P. Moore, Lutterworth Press, 1977.
8. *The history of life*, by A.L. McAlester, Prentice-Hall, 1977 (this covers life on the Earth only).
9. *The inaccessible Earth*, by G.C. Brown and A.E. Mussett, George Allen & Unwin, 1981 (mainly on the Earth's interior).
*10. *Introduction to comets*, by J.C. Brandt and R.D. Chapman, Cambridge University Press, 1981.
11. *Jupiter*, by G. Hunt and P. Moore, Mitchell Beazley, 1981.
12. *Lunar science: a post-Apollo view*, by S.R. Taylor, Pergamon Press, 1975.
13. *The Moon*, by P. Moore, Mitchell Beazley, 1981.
14. *The Moon — our sister planet*, by P. Cadogan, Cambridge University Press, 1981.
*15. *The nature and origin of meteorites*, by D.W. Sears, Adam Hilger, 1978.
*16. *The origin of life on the Earth*, by S.L. Miller and L.E. Orgel, Prentice-Hall, 1974.
*17. *The planet Pluto*, by A.J. Whyte, Pergamon Press, 1980.
18. *The surface of Mars*, by M.H. Carr, Yale University Press, 1981.
19. *Understanding climatic change*, National Academy of Sciences, USA, 1975 (this covers the Earth's climate only).

Life in the Universe

*1. *A discussion on the recognition of alien life*, papers given at a 1974 Royal Society discussion meeting, The Royal Society, London, 1975.
*2. *Extraterrestrials: where are they?*, ed. by H.H. Hart and B. Zuckerman, Pergamon Press, 1982.
3. *Lifecloud*, by F. Hoyle and N.C. Wickramasinghe, J.M. Dent & Sons, 1978.
*4. *Life in the Universe*, ed. by J. Billingham, The MIT Press, 1981.
5. *The quest for extraterrestrial life*, ed. by D. Goldsmith, University Science Books, 1980 (contains a good deal of historical papers).

Periodicals

1. *Astronomy* (monthly, devoted to articles and news items on astronomy).
2. *New Scientist* (weekly, contains articles and news items on astronomy).
3. *Scientific American* (monthly, contains articles on astronomy).

4. *Sky and Telescope* (monthly, devoted to articles and news items on astronomy).

*5. *The Observatory* (bi-monthly, devoted to articles and reviews on astronomy).

6. *Yearbook of Astronomy*, pub. by Sidgwick & Jackson (annual, devoted to calendars of astronomical events, star charts and articles on astronomy).

INDEX